MINING IN
THE HIMALAYAS
An Integrated Strategy

MINING IN THE HIMALAYAS
An Integrated Strategy

A. K. Soni

CRC Press
Taylor & Francis Group
Boca Raton London New York

CRC Press is an imprint of the
Taylor & Francis Group, an **informa** business

CRC Press
Taylor & Francis Group
6000 Broken Sound Parkway NW, Suite 300
Boca Raton, FL 33487-2742

First issued in paperback 2020

© 2017 by Taylor & Francis Group, LLC
CRC Press is an imprint of Taylor & Francis Group, an Informa business

No claim to original U.S. Government works

ISBN 13:978-0-367-57447-5 (pbk)
ISBN 13: 978-1-4987-6234-2 (hbk)

Visit the Taylor & Francis Web site at
http://www.taylorandfrancis.com

and the CRC Press Web site at
http://www.crcpress.com

I dedicate this book to the victims of the April 25, 2015,

earthquake that gripped the Nepali Himalayas.

Let us dive into a good book to conserve nature!

Contents

Preface

Many technical papers, articles and excellent books have been written on the Himalayas, and literature is abound on the Himalayan environment and related issues; however, very few are available on *Himalayan mining*. This brings forth the question 'Why?' And the answer is ... *Mining in the Himalayas* is a very topical and contentious subject, which may raise eyebrows of a number of persons who have very little knowledge and interest in mining. I realized that there is a dire need for the analysis on the subject; hence, I penned this book. Thus, this publication focuses on some of the earlier gaps in mining and mine technology areas especially:

- *How to make use of existing mining practices so that best results are achieved (Best Practice Mining)*
- *Some new practical ideas and eco-friendly solutions for the Himalayan mining*
- *How quantification of environmental degradation of fragile Himalayan mining areas is done so that optimum results are achieved (Environment Degradation Index)*

The Himalayas, the most prominent mountain system of the world, is a source of sustenance to the social, cultural and economic development of the diverse Himalayan population across different countries extending in India, China, Afghanistan, Pakistan, Nepal, Myanmar, Bhutan and Bangladesh (Hindukush Himalayas). The entire extrapeninsular region of India (northern and north eastern parts) is covered with the cordillera of the Himalayas. In the past, the neglect of mountain areas by policy makers has resulted in the slack infrastructure development of such areas despite the rich natural resources they possess. Limited scientific efforts have been made in the past to harness 'minerals' which are found in the mountainous Himalayan regions/geosettings. As a result, *unscientific small-scale mining* have grown leaps and bounds in the region and took a heavy toll on the *ecosystem*. This needs an *integrated* and *holistic* approach to deal with the resources management which also has a great bearing on the plains. In view of its rugged topography, harsh climatic conditions and environmental sensitivity, such approaches can repair the Himalayan eco-system adequately which had been ravaged in the recent past due to human interventions.

With 8 well-described chapters, 212+ text pages and comprehensive bibliography, this book makes a concerted effort to explain how mining in the Himalayas is practiced and what is to be done in the future? Limestone is found in abundance in the Indian Himalayas; hence, its details are explained in the Himalayan mining practices, and discussions are made on the repercussions of limestone mining with respect to the environment. Critiques and lacunas of mining and environmental practices are discussed at length, and the case studies of limestone, magnesite and soapstone are described to illustrate the existing mining practices. Hill slope management and best mining practices can be dovetailed with the social fabric of the hilly Himalayas, and scientific analysis can be made possible. Three decades of rich experience of the author explains how advance developmental planning can assist in project efficiency.

The Himalayan region is characterized by inaccessibility, fragility, marginality and diversity. The mountains provide a niche of opportunities for development in which they have comparative advantages such as hydropower, tourism and horticulture. Considering these characteristics/features, a technical explanation has been given for the reader to understand the *eco-friendly perspective of Himalayan mining.* This is described in the form of a question and answer for the debatable topic of 'mining in the Himalayas'.

This book contains some new, practical and eco-friendly solutions/ideas for the Himalayas. These new features are suggested for the readers and industrial organizations through eight cases, namely *underground mining of limestone; cluster mining concept for small-scale mining; continuous miner; solution mining of rock salt; innovative transportation system in use at Himalayan mine; slope hoisting system, high angle conveying for ROM hoisting from valleys and low cost slope stabilization measures.* Environmental degradation index (EDI) for application especially in fragile areas especially is yet another new feature of this book. For the development of this analytical tool, that is EDI, the research input based on field studies is taken and incorporated in this book. Environmental parameters such as air, water and land component of the environment parameters are incorporated in the EDI to know the status of the environment and the cost involved in environmental protection at a particular site selected for the study.

The sincere efforts made by the author thus suggest solutions and develop concepts for mineral resource management. The attempt so made has led to the *integrated development approach* for the mineral-bearing areas of the Himalayas. This has been done, keeping in mind the inherent damages and the mining causes to the ecosystem. 'Best Management Practices' (BMP) in mining, which is an improvement of existing practices; for example *overland belt conveyor system* is the most economical, practically possible and eco-friendly solution for ROM transportation in the Himalayas for the *Mangu* and *Pati* sub-block of *Kashlog mine.* These best practices are subdivided into two parts as 'mining problems' and 'environmental problems'. A scientific approach is adopted to deal with the potential and environmental-friendly engineering solutions or measures for artisanal small-scale mining sector. It is suggested that their successful implementation will protect the fragile environment of the Himalayas. In the seventh chapter of the book, an emphasis is given for *environment-oriented development,* which can provide the best possible solution for the affected environment. *Road map* and *lessons learned* also form a part of this chapter. In this way, a holistic approach has been adopted for mining in the hilly areas.

The formulation of *integrated strategy* on watershed basis for the development and exploitation of mineral resources of the Himalayas is fully explained as this strategy is best suited, eco-friendly and practically applicable and opens a new vista for the sustainable development of the Himalayan region as a whole.

It is hoped that the book will play its role as a purveyor of scholarly scientific information by focusing on the environmental issues of mining areas which are specific to the Himalayas. By utilizing the suggested concepts, maximum possible benefits can be harnessed and mining can continue in an eco-friendly manner. Thus, *environment improvements in mining areas* are the key benefits of reading this book. The work described would be helpful to get the answer to all those who think that mining in the Himalayas is a very difficult task. I conclude this preface with this optimistic note that eco-friendly mining is possible had we avoid shortcuts, adopt scientific approach of mining and make use of mineral conservation practices, thereby paving the way for future generations.

Dr. Abhay Kumar Soni

Acknowledgements

It took me about 15 years since 1999 to analyze, complete and portray this research work on Himalayan mining which forms the content of this book. Inclusion of practical ideas in this piece of work is an outcome of the experience I gained while working as a researcher at the Central Institute of Mining and Fuel Research (CIMFR; formerly known as Central Mining Research Institute [CMRI]). I express my thanks to all the authorities and staff of CSIR-CIMFR for their periodical help and support.

I am extremely thankful to all my colleagues at Nagpur and Roorkee for helping me, directly and indirectly, in generating ideas expressed in this book. I acknowledge all my colleagues at Nagpur whose names I cannot mention one by one. I heartily acknowledge Dr. R.K. Goel, chief scientist, CSIR-CIMFR, Roorkee, for his encouragement and motivating me in bringing my research work into a book form. I also acknowledge my thanks to Sri Y. Raghvendra Rao, former vice president (Mines & Geology), Ambuja Cement; Sri Ankur Agarwal, senior GM (Mines), Ambuja Cement, Darlaghat (H.P.); Sri Y.K. Sharma, managing director, Almora Magnesite Limited, Almora; Shri S.K. Sinha, Jaypee Cement; and Sri D.K. Sharma, secretary, Department of Mines and Geology, Government of Sikkim, who shared technical information about the mines which lie in their jurisdiction in their contribution and for their timely help. Sri Rajneesh Sharma and Shri Arun Sharma, state geologists, Department of Industries, Shimla, (H.P.); Sri Rajendra Tewari, general manager, Jai Singh Thakur & Sons, Paonta Sahib (H.P.); and Shri S. Bhardwaj, mining Officer, Department of Industries, Government of Himachal Pradesh, Nahan (H.P.), are thankfully acknowledged for updating information about Sirmour mining and mining in Himachal. I also express my deep sense of gratitude to my senior colleagues and to my PhD supervisors Prof. N.C. Saxena and the (late) Dr. A.K. Dube who encouraged me to research on mining in the Himalayas.

I feel obliged to the Taylor & Francis Group and in particular to Dr. Gagandeep Singh, editorial manager, Engineering/Environmental Sciences at CRC Press, and his team for taking all the pains to publish this book with professionalism and with deep interest. Their long-standing experience has shaped the book in its present form, and I thankfully acknowledge it.

Document of this shape and size always entails fruitful discussions among its stakeholders or those who are involved with the subject. I thankfully acknowledge all those who contributed through fruitful discussions during the course of the preparation of the manuscript of this book. I acknowledge the help of Dr. P. Nema, former director grade scientist, NEERI, Nagpur, for carefully reading the manuscript and giving his valuable and experienced suggestions. Mrs. Anjali Jaipilley and Shri Ranjit Mandal's help in the preparation of the figures, drawings and photographs, and improvement thereof are duly acknowledged. Last but not least, I thank all the authors and publishers who permitted me to use their published papers, documents, tables, figures, etc., in my book and all the expert reviewers for giving constructive suggestions.

I would be failing in my duties if I do not remember and place on record my appreciation to my family members: my wife Archana, my daughter Niharika, my son Navneet and my mother, for their constant moral support and encouragement during the entire journey of writing this book.

Dr. Abhay Kumar Soni
Nagpur, India

Author

Dr. Abhay Kumar Soni received his bachelor's degree in mining engineering from the National Institute of Technology, Raipur, Chhattisgarh, India, in 1983 and his MS in science and technology from BITS, Pilani, India, in 1991. He has completed his Ph.D. in environment science and engineering from the Indian School of Mines (now IIT), Dhanbad, India, in 1998 and gained more than 30 years of working experience in the Indian mining industry. As a research scientist and a technical administrator, he possesses an in-depth knowledge of the field, teaching and research. Dr. Soni has more than 100 technical publications on environmental-related topics to his credit which are published in national and international journals, monographs, conference proceedings, workshops, etc. He has authored technical papers in English as well as in Hindi and received honours and awards at both national and international levels, such as being a member of the international advisory board for the *Journal of Mine Water and Environment*, published by Springer Verlag, and receiving awards for best technical papers by the ISRMTT (2013), IGS (2003) and IMEJ (2005).

Dr. Soni is a member of important committees and an expert member/evaluator of and responsible for a number of noted professional assignments. He is actively associated with technical and professional organizations and societies, namely the Mining Engineers Association of India (MEAI), Institution of Engineers (India) and International Mine Water Association (IMWA). Since 2009, he has been the scientist-in-charge and head of CSIR – Central Institute of Mining and Fuel Research, Nagpur, India.

Abbreviations

ACC	Associated Cement Company Limited
ACL	Ambuja Cements Limited. (Himachal Unit of Ambuja Cement, formerly HACL)
BIS	Bureau of Indian Standard (Formerly ISI)
CCI	Cement Corporation of India Limited
CIMFR	Central Institute of Mining and Fuel Research (CIMFR) (Formerly, Central Mining Research Institute [CMRI])
CMRI	Central Mining Research Institute
CPCB	Central Pollution Control Board
EFA	Ecologically fragile areas
EIA	Environment impact assessment
EMP	Environment management plan
EPA	Environment Protection Act, 1986
EPR	Environment Protection Rules, 1986
GSI	Geological Survey of India
IBM	Indian Bureau of Mines
ICIMOD	International Centre for Integrated Mountain Development, Kathmandu (Nepal)
MECL	Mineral Exploration Corporation Limited
MOEF	Ministry of Environment, Forest and Wildlife, Government of India
MRL	Mean reduced level (elevation/level above ground level)
MT	Million tonnes.
NMDC	National Mineral Development Corporation
ROM	Run of mine
SSM	Small-scale mines/small-scale mining
tpa	Tonnes per annum
tpd	Tonnes per day
tpy	Tonnes per year
σ	Overall stable pit slope

1

The Indian Himalayas: Mining and Mineral Potentiality

India is a vast country and rich in mineral resources. These resources are very important to the national economy and constitute the main building blocks of development. In India, 87 minerals are produced (4 fuel minerals, 3 atomic minerals, 16 non-metallic minerals, 10 metallic minerals and 54 minor minerals) from small, large and very large deposits and are being exploited for commercial purposes (IBM, 2015). Since mineral resources are site-specific and non-renewable and have uneven regional distribution, some regions have abundance of one mineral type, whereas others are deficient in them. A similar trend has been found in the context of the Himalayas as well, which is said to be a storehouse of a variety of mineral deposits in the Indian peninsula.

In the past, it was realized by the policy makers that the lack of authentic data (reserves, geology) and understanding of natural ecological processes of the mountains and their environmental geosetting hit the mountain people hard and prevented the exploitation of resources. To keep pace with the growing mineral demand and regional development, Indian think tanks started considering the formulation of policies that were specific and beneficial to that region. Accordingly, a policy framework that enables infrastructure development for the hill region, in particular for the Himalayas, along with improved management of the mountain's natural resources, was mooted. This has been augmented at the time when the largest and most populous state of India (Uttar Pradesh, UP hereafter) was divided into two parts according to their geographical features, namely Uttarakhand (UK, all hilly areas) and UP (all non-hilly/plain areas).

This chapter provides a precise and at-a-glance description of the Himalayas, the Himalayan region and its characteristics. An account of the overall mineral potential of the Indian Himalayan region is presented, which otherwise remained scattered and obscure. A short explanatory note giving analysis and discussion on the possibility of their commercial exploitation is given with a view to understanding their utility for the country and industry. Mining practices and legal obstacles faced by the mining community form part of the chapter to know the reality. It is felt that any 'resource use practices' or 'development interventions' with reference to the Himalayas must draw the attention of mountain specificities as well, which include inaccessibility, marginality, fragility, diversity, comparative advantages (niche) and adaptation experiences (ICIMOD, 1995).

1.1 Introduction

It is well recorded and known that nature has endowed India with rich mineral resources and a vast variety of land forms, spread over the diverse geographic and climatic conditions: that is, snowy peaks of the Himalayas, fragile foothills, arid deserts and semiarid areas, vast coast lines, plain lands and dense forests with a rich biological diversity.

The Himalayas, the most prominent mountain chain of the world, are a cradle of ancient civilization. It is a source of sustenance for the social, cultural and economic development of the entire extra-peninsular region of India and home to about 9% of India's population (Figure 1.1). The Himalayan ecosystem, in view of its mountainous topography, rugged terrain, sociocultural diversity and environmental sensitivity, is a region quite different from others in terms of its resource management. These resources could be in the form of water, air, land or minerals. All these vital resources are the important constituents of the environment, and therefore, emphasis has been placed on each of them, giving due importance to both ecology and the environment. However, the core/focus is on the mineral and mining segment.

The Himalayas that we analyze here are a dynamic hill resource system and its management needs site-specific solutions for sustainable development. In view of adversities, complex field conditions and industrial attempt against nature, the management of resources (including mineral resources) needs a

FIGURE 1.1
The Indian Himalayan region.

holistic approach for scientific solutions. Since mineral resource exploitation in the Himalayas is an intricate and debatable subject, appropriate answers based on scientific facts and evaluations are the best and urgently required.

1.2 The Himalayas

The Himalayas (हिम+आलय meaning 'snow dwellings' in Hindi or its literal meaning in Sanskrit is 'abode of snow') are a mountain range in South Asia, which separates the Indo-Gangetic Plain (India) from the Tibetan Plateau (China). The Himalayas span many countries, namely India, Nepal, Bhutan, China (Tibet) and Pakistan (Figure 1.2), with the first three countries having sovereignty over most of the Himalayan range.

In entirety, *the Himalayas* extending in different Asian countries are commonly referred to as the Hindu Kush Himalayas (Box 1.1). Afghanistan

FIGURE 1.2
The Hindu Kush Himalayas extended in Asian continent.

BOX 1.1   HINDU KUSH HIMALAYAS

The word 'Hindu Kush' consists of two words: Hindu and Kush. The word 'Hindu' originally refers to any inhabitant of the Indian subcontinent or followers of the Hindu religion extended mostly in India, Nepal, Bhutan, Bangladesh, Pakistan, Mauritius and Southeast Asia. However, the interpretation of word 'Kush' is contentious. In the modern Persian language, the word 'Kuh' means a mountain and the word 'Kush' is derived from the verb 'Kushtan' meaning to defeat, kill or subdue. Some refer the word 'Hindu Kush' as 'Hindoo-Kho' or 'Hindoo Kush' also. However, etymologically, numerous possibilities for its origin have been put forward, and the popular word 'Hindu Kush' is now accepted. The word 'Hindu Kush' in Sanskrit is also known as 'Pariyatra Parvata' (पारियात्र पर्वत) and derived from 'Hindu-Kuh', meaning the 'mountain of Hind'.

It is a 800–966 km (lateral) × 240 km (median north–south measurement) mountain range that stretches between central Afghanistan and northern Pakistan. It is a western sub-range of the Himalayas. It divides the valley of the Amu Darya to the north from the Indus River Valley to the south. Only about 600 km of the Hindu Kush system is called the 'Hindu Kush mountains', and the rest of the system consists of numerous smaller mountain ranges, for example *Safed Koh* and *Suleiman Range*. The highest point in the Hindu Kush is the *Tirich Mir* (altitude: 7708 m) in the Chitral district of Khyber Pakhtunkhwa, Pakistan. To the east of the Hindu Kush lies the Pamir Range near the point where the borders of China, Pakistan and Afghanistan meet, after which it runs southwest through Pakistan and into Afghanistan, finally merging into minor ranges in western Afghanistan. The mountain range separates Central Asia from South Asia.

The Himalayas/Hindu Kush Himalayas have profoundly shaped the cultures of South Asia and has tremendous international and national importance. The mountains of the Hindu Kush system diminish in height as they stretch westwards. Towards the middle, near Kabul, they extend from 4500 to 6000 m and in the west they attain heights of 3500 to 4000 m. The average altitude of the Hindu Kush is 4500 m.

The Himalayan mountains are of historical significance in the Indian subcontinent and China. There has been a military presence in the mountains for quite some time. The great game involving military, intelligence, espionage and so on from across the countries' boundaries was operative in the areas of the Hindu Kush. Before the Christian era, and afterwards, there was an intimate connection between the Kabul Valley and India. All the passes of the Hindu Kush region used by the travellers from the north were helpful in the development of civilization and religion. The great Himalayan range was the boundary and barrier between the countries for time immemorial.

Source: From various websites.

(western subrange of the Himalayas in central Afghanistan/Pamir Range) and Myanmar (Burma) are also part of the Hindu Kush Himalayas. The Indian subcontinent, divided geographically into the peninsular, extra-peninsular and the Indo-Gangetic Plain, encompasses the lofty hill ranges of the Himalaya (northern and northeastern part), which is mainly a part of the extra-peninsular region. The Chittagong Hill Tracts of Bangladesh are not exactly the Himalayas but are covered in the Hindu Kush region.

The Himalayan range of mountains includes 9 of the 10 highest peaks on Earth, including the world's highest, the Mount Everest. Many Himalayan peaks are sacred in both Buddhist and Hindu religions. Three of the world's major rivers, namely the Indus, the Ganges and the Brahmaputra, rise in the Himalayas. While the Indus and the Brahmaputra rise near the Mount Kailash in Tibet, the Ganges rises in the Indian state of Uttarakhand (UK). Their combined drainage basin is home to some 600 million people.

The Himalayas were lifted by the collision of the Indian tectonic plate with the Eurasian Plate, and the Himalayan range runs from northwest to southeast in a 2400 km long arc. The range varies in width from 400 km in the west to 150 km in the east.

In India, the Himalayas are divided from south to north into the sub-Himalayas or Shiwaliks or Siwaliks (up to 1000 m altitude), lesser Himalayas or middle range Himalayas (2000–3000 m altitude), great Himalayas and trans-Himalayas. From west to east, the Himalayas in the Indian territory (Figure 1.3) are sub-divided region wise and categorized into the following:

- *J&K Himalayas* or the *northwestern Himalayas* in the state of Jammu and Kashmir
- *Shimla Himalayas* or *HP Himalayas* in the state of Himachal Pradesh (HP)
- *Garhwal Himalayas* extending in the hill districts of Garhwal (previously also referred to as UP Himalaya)
- *Kumaon Himalayas* in the Kumaon division of UK
- *Northeastern Himalayas* extending in the states of Arunachal Pradesh, Sikkim and Assam

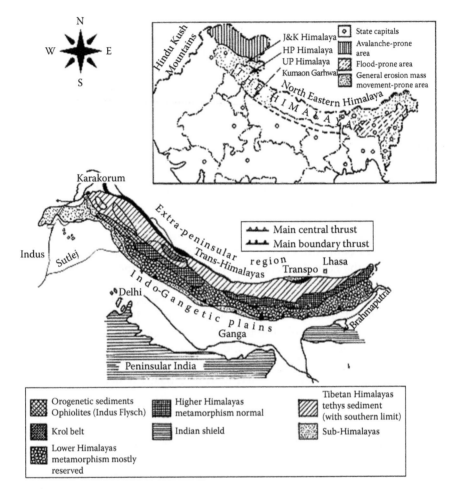

FIGURE 1.3
Sub-regions of the Himalayas in the Indian territory.

Thus, according to India's country boundaries, the Himalayas are bordered on the north by the Tibetan Plateau, on the south by the Indo-Gangetic Plain, on the northwest by the Karakoram and Hindu Kush ranges and on the east by the Indian states of Assam and Arunachal Pradesh.

1.2.1 Typical Characteristics of the Himalayas and Their Fragility

The typical characteristics of the Himalayas make them a different region from the others. Some of these significant characteristics are the fragility, seismicity and rugged topography. The Himalayas are also characterized by mountain specificities, for example inaccessibility, marginality (soil and lands are marginal in hills) and physical/biological/social/cultural/ economic diversity. The Himalayas have comparative advantages and varied adaptation experiences over the plains. They incorporate some natural opportunities too, which prevent their overexploitation and encourage traditional skills. All these typical characteristics have a bearing on the resource management of this hilly region.

The Himalayas are a highly seismic zone and comprises a tectonically active region. The occurrence of earthquakes and tectonic activities, a basic reason for the rise of the Himalayas, is quite predominant in these hilly areas. Recent tectonic models from structural, metamorphic, geochemical, geochronological, geophysical and other perspectives (Mukherjee et al., 2015) have postulated new and modified theories about the Himalayan orogeny in last 15 years (Mukherjee et al., 2013).

In the seismic map of India, the Himalayan region falls under 'Category V', which means a 'seismically active area' (Figure 1.4). Therefore, any industrial or developmental project should consider such seismic forces in their design procedures and take into account a higher factor of safety. Extra precautions are therefore needed for such areas.

The fragility of the Himalayan ecosystem in India (MOEF, 1990) is mainly defined and related to its

- Weak and younger geological formations
- Arrested (vegetation) succession
- Desertification

Weak and younger geological formations under the influence of human impact and mountainous topography lead to frequent landslides, which are quite common in the lower and central parts of the Himalayas. Because of such ground conditions, mature forests and vegetation have not developed and the conditions have led to continuous cutting of gorges by rivers in the deep valleys, causing increase in the degree of steepness and adding to the problem of hill stability.

Arrested succession is a common feature in the Himalayan region. Exotic weeds (*Lantana camara, Eupatorium* spp., *Mikania micrantha*) are widespread in the northwestern and northeastern Himalayas. As a consequence, there is continuing decimation of the germplasms of larger shrubs and trees. Another example is the large-scale transformation of landscape for chir-pine plantation, resulting in drastic alteration in the physicochemical properties of soils and displacement of mature oak forests.

Desertification is another feature of the Himalayas. The rain forests of the Himalayas, particularly in northeastern India, are developed on extremely poor, nutrient-deficient soils. These forests are sustained because of the thick roots developed over the soil surface, which help in rapid recycling of nutrients. The nutrients released from the decomposing leaf litter are

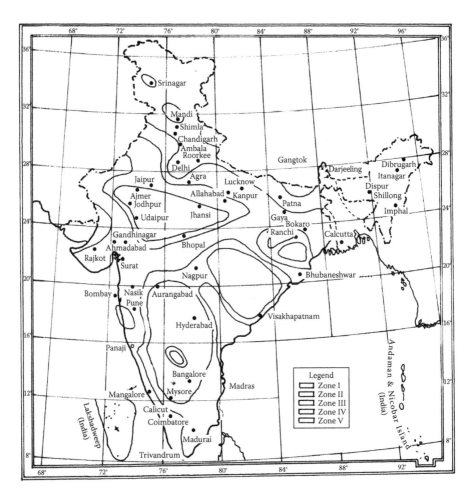

FIGURE 1.4
Tectonic map of India.

short-circuited by the surface root mat and put back into the living biomass even before it can enter the soil. Felling of trees upsets this delicate balance in nutrient cycling, with the result that the forest ecosystem fails to recover and the land is often desertified. This has occurred and been recorded in the northeastern Himalayas.

It should be noticed here that to lay down a set of uniform parameters for all fragile hill ecosystems is a difficult task due to the fact that their environmental setting, climatic conditions and geological fabric are different though they may have similar rugged terrain or topography. Considering these facts, different countries have laid down different guidelines for declaring their hill areas as 'fragile areas'. Thus, it is apparent that the fragility parameters described earlier are true for the Indian Himalayas only.

1.3 Hill Areas Comparable to the Himalaya

The Himalayas can be compared on a continental scale and a local national scale. Three mountain chains of the world (Figure 1.5) are comparable to the Himalayas.

- *The Alps* in Europe
- *The Andes* in South America
- *The Rockies* in North America

In order to make a comparison of the Himalayas with other ecologically fragile hill areas (EFAs) of the rest of the world, two important points need to be considered: (1) its geological setting and topography and (2) the environmental sensitivity, often referred to as fragility. The Appalachians and Ural Mountains (Figure 1.5) are not comparable with the Himalayas in terms of the fragility parameters mentioned earlier.

Geologically, the Himalayas have been considered as the *metallogenic terra incognita* (Dar, 1968). Indian geologists have found many similarities between the Himalayas and other contemporary mountain ranges, for example the Alps in Europe.

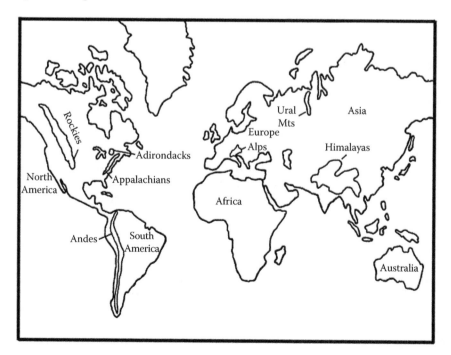

FIGURE 1.5
Mountains contemporary to the Himalayas.

While exploring minerals in the Himalayas, one also has to look into the experiences gained in a similar contemporary system, such as the Alps system, and its ramifications within the Himalayan context. Both the Alps and the Himalayas are of Mesozoic–Cenozoic age evolved by a simultaneous mountain-building process, though the latter is younger in formation. Extensive studies carried out on these two mountain chains have shown that in the Alps/Alpine region most of the mineralization is associated with alternating succession of volcanic and sedimentary suite of rocks. The orogenic and metallogenic processes of mineral formation in the Himalayas and the Alps are very similar. Stratigraphically, the Alpine rocks are known to house metallic and non-metallic minerals in large quantities at staggered places (Nair and Mithal, 1976).

In India, the *Aravallis*, located in the mineral-rich state of Rajasthan, is also an EFA comparable to the Himalayas. The rock formations in the Aravallis are stable and subject to less tectonic disturbance as compared to the Himalayan rock formations, but these mountains are comparable to the Himalayas in terms of 'fragility'. In 1992, the federal government of India prohibited the continuance of all new mining operations including renewals of mining leases, existing mining leases in sanctuaries/national parks and areas covered under Project Tiger. Mining without the competent authority's permission is declared illegal. Prior permission for activities in such areas is given by the central Ministry of Environment and Forests (MOEF), Government of India vide notification S.O. 319(E), No. 17/1/91-PL/IA dated 7 May 1992 (www.moef.nic.in/legis/eia).

1.4 Mineral Potential of the Indian Himalayas: An Account

The scrutiny of national mineral inventory records prepared periodically by the Indian Bureau of Mines (IBM) indicates that no separate estimation of mineral reserves for the Himalayas was done in the past. These are estimated as part of state mineral reserves records, which are reported by IBM (2015).

Deposits of fuel minerals as well as metallic and non-metallic minerals have been reported in the Himalayan region since the last century, and sporadic mining activities all over the Himalayas have been going on for many years. Based on a literature review, a brief account of the minerals of the Indian Himalayas and their economic importance from the commercial exploitation view point has been prepared, which is given in Table 1.1 (Soni, 1997). The deposits of copper, lead and zinc are reported from the eastern and western part of the Indian Himalayas, which are mainly base metals. Base metal deposit of Rangpo, Sikkim, is an example

TABLE 1.1

Location-Wise Minerals of the Indian Himalayas and Their Economic Status (in Two Sections)

Mineral	State	Location	Economical Status	Remarks
Section A: Minerals of Economic Importance from Exploitation Point of View				
Coal	Sikkim	Namchi area (South Sikkim)	1	High ash content.
	Sikkim	Reshi (West Sikkim)	1	Gondwana-type coal.
	Arunachal Pradesh	Kameng, Subansiri, Siang and Tirap	3	Large-scale mining operation is difficult to perform due to geological conditions.
	Assam	United Mikirs, North Cachar Hills, Sibsagar and Lakhimpur	2 and 3	Tertiary coal friable in nature with high sulphur content.
	J&K	Nirpur, Poonch, Rajauri and Udhampur of the Jammu region (Kalakot, Metka and Mahogota coalfields under extraction)	2	The coal in the Jammu region is of anthracitic variety.
	J&K	Lignite seams of Kashmir Valley (Karewa sediments)	3	Reserves are limited in quality.
Rock salt	HP	Mandi	2	Guma salt mines and Drang mine of Hindustan Salt Ltd. Small-scale manual mining under operation (refer Chapter 4 for other details). Rock salt is impure, hence require beneficiation.
Magnesite	UK	Almora and Pithoragarh, UK	2	Associated rock types are dolomite and talc
	UK	Chamoli, UK	2	Economical deposit exploitable by small-scale mines
	HP	Chamba, UK	1	–
	J&K	Udhampur, UK	1	–
Base metals	Sikkim	Bhotang, Rangpo and Dikchu area (east Sikkim) (Cu–Pb–Zn)	2	Mining operation done earlier but discontinued since 2006
	J&K	Anantnag, Rajauri, Udhampur, Doda and Baramula	3	Only occurrences (economical exploitation needing scientific evaluation)

(Continued)

TABLE 1.1 (*Continued*)

Location-Wise Minerals of the Indian Himalayas and Their Economic Status
(in Two Sections)

Mineral	State	Location	Economical Status	Remarks
	UK	Pithoragarh and Almora	1	–
Gypsum	J&K	Doda and Baramula (Ramban, Asar, Batote and Baniyar areas)	2	–
	UK	Dehradun and Tehri Garhwal	2	–
	MP	Chamba and Sirmour	2	–
Slate	HP	Dharamshala (Ghanyara and Dari Panchayat)	2	Under extraction by local villages on small scale under the administration of Panchayat
Phosphorites	UK	Mussoorie (Doon Valley) phosphorites	2	Underground mining being carried out by PPCL
	UK	Tehri Garhwal phosphorites	1	Small deposits (extension of Mussoorie Phosphorite deposit)
Graphites	Arunachal Pradesh	Lohit and Subansiri	1	–
		Baramula	1	–
Limestone	Assam	North Cachar Hills; United Mikirs	2	Available in abundance in all parts of the lower Himalaya; Under commercial exploitation by small-scale mines as well as large-scale mines
	HP	Kangra, Kulu, Mahasu, Mandi, Sirmour, Solan and Bilaspur	2	
	UK	Dehradun (Mussoorie), Tehri Garhwal, Almora Nainital and Pithoragarh	2	
	J&K	Anantnag, Baramula, Kathua, Punch, Rajauri and Udhampur	2	
	Arunachal Pradesh	Lohit and Subansiri	1	
Dolomites	Arunachal Pradesh	Kameng, Arunachal Pradesh	1	–
Barytes	HP	Sirmour	2	Being extracted by small-scale mining operation
	UK	Doon Valley	3	Only occurrences

(Continued)

TABLE 1.1 (*Continued*)

Location-Wise Minerals of the Indian Himalayas and Their Economic Status
(in Two Sections)

Mineral	State	Location	Economical Status	Remarks
	J&K	Udhampur, J&K	1	–
Talc, soapstone, steatite	UK	Almora, Chamoli and Pithoragarh, UK	?	Under extraction at some places

Section B: Minerals of Occurrence Nature/Trace Values (of Less Economic Importance from the Exploitation Angle)

- Scheelite is reported to occur in the Ranikhet area of Almora district.
- Tungsten exists in the Chamoli district of Uttarakhand and 0.13 million tons reserves at Karauli and Alai in Almora district.
- Gold is found in Garhwal region and Siwalik belt.
- Sapphire occurs at Sunsan, Paddar and Ramshy in Doda district (lenticles in kaolinized pegmatites and the associated rocks are crystalline limestone).
- In the middle Shiwalik of the Sulaiman Range, strata-bound uranium mineralization has been reported from sand stone rock type. Uranium mineralization from the lower Shiwalik has also been reported (limited in reserve is therefore not economical for commercial exploitation).
- Boron conditional (submarginal) resources of 74,204 tons have been estimated in the Leh district of J&K.
- Bentonite is an important naturally occurring clay of great commercial importance. It is commonly known as bleaching clay and is formed by the alteration of volcanic ash or tuff. It is reported to occur in the Jammu region.
- Occurrences of potash and potash salts have been reported from Tsokar Lake (Leh district) and J&K but are of not much economic significance.
- Petroleum and natural gas are found in Assam and Arunachal Pradesh (some areas of the Himalaya are promising from the view point of these resources).

Source: Soni, A.K., Integrated strategy for development and exploitation of mineral resources of ecologically fragile area, PhD thesis (Unpublished), Indian School of Mines (ISM), Dhanbad, India, 1997, p. 238 (compiled from various published and unpublished official documents of GSI, MECL, IBM, and others).
Note: 1, not known; 2, economical; 3, uneconomical.
Abbreviations: HP, Himachal Pradesh; UK, Uttarakhand; AP, Arunachal Pradesh; J&K, Jammu and Kashmir (all these are states of India).

of proven and economically extractable deposit containing Cu/Pb and Zn. Geological explorations point out the following expectations for the present and future:

- Low-grade copper ore bodies in massive igneous rocks of the Ladakh and Kashmir Valleys
- Lead–zinc deposits in the middle Palaeozoic limestones in the state of UK (erstwhile western UP)

- Copper deposits in sandstone sequences of Upper Palaeozoic lime-stones in the state of UK
- The sub-Himalayan rocks for placer and palaeo-placer deposits of economic importance
- Uranium mineralization in Siwalik sediments
- Hydrocarbons and coal in the sub-Himalayas and the lesser Himalayas (not yet fully explored)
- The lesser Himalayas for large high-volume, low-cost deposits of limestone, dolomite, magnesite, phosphorite, graphite, gypsum and rock salt
- The lesser Himalayas for small- to medium-size polymetallic sulphide deposits
  - Precious metallic ores like gold ore and numerous rare gems (ruby, sapphire, asphaltic, etc.) as trace minerals (largely unorganized and small deposits at scattered places)
- Bedded sulphides in the lower Palaeozoic sequence (Tethyan Himalaya)
- The lesser Himalayas and Central Crystalline Zone for tin–tungsten mineralization, sulphide mineralization and precious/semi-precious stones
- The Indus and Shyok belts and trans-Himalayan Plutons of Ladakh, Karakoram and Mishmi for low-volume, high-cost metalliferous deposits and precious/semi-precious stones
- *Metallic minerals*: In the Alaknanda Valley, Garhwal (UK), Bhowali–Bhimtal–Ladhiya Valleys, Kumaon (HP) and the Garsch–Parvati Valley of Kullu (HP), there are evidences of thick piles of volcano-sedimentary suite of rocks of early Palaeozoic age. These valleys of the Himalayan region offer tremendous scope for detailed exploration to find metallic minerals. (The scenario is similar to the Alpine region where highly mineralized beds exist.)
- *Limestone*: High-quality limestone is known to occur at number of places in the entire Himalayan region in huge quantities. The Krol formations contain mainly limestone, dolomite, dolomitic lime-stone, sandstone, clay, shale and quartzite deposits besides other minerals such as phosphorite, which are found in small/limited quantity only

The 'Krol Belt' (geological name of a belt) extending in the Himalayas over a length of more than 350 km and named after the 'Krol Peak', which has an altitude of 2278 m above mean sea level, constitutes a chain of the Himalayan ranges stretching from the Kunihar Valley in the north to the Tons River in the southeast in the HP Himalaya (Shimla and others). It continues to the east up to Nainital in the Kumaon Himalayas. The Krol formations are considered as one continuous horizon but occurring as two distinct structural belts (Auden, 1934). The Krol belt comprises of six prominent mountain ranges and contains industrial-grade limestone for various uses (Bhargava, 1976).

On the basis of the investigations done by the principal geological exploration and reserve estimation agencies, namely the Geological Survey of India (GSI), Mineral Exploration Corporation Limited (MECL) and other state mining and geology agencies, some deposits have been proven as economically exploitable for commercial purposes (Table 1.2).

The tectonic settings of the Himalayas, when compared with similar hill system of the world, indicate that metals in economic abundance could be present in this huge mountain range. But according to Pachauri (1992), different tectonic settings of the Himalayas created by the collision of the Indian plate and the Asian Plate have produced zones of potential mineralization, but overall the Himalayas are not highly mineralized. One probable reason for the lack of significant mineralization is the young age of the Himalayan mountain system and the occurrence of significant erosion (Pachauri, 1992).

In terms of mineral deposits and tectonic settings, the Himalayas can be broken up into the following north to south tectonic settings: precollision basins, ophiolitic suite of rocks, hinterland region (north of the Indo-Tibetan suture zone [ITSZ]), barren hinterland basins, foreland thrust belt zone, foreland basin and sedimentary sequences in the Klippe zone (Figure 1.6). Ores and minerals found or likely to be present in these zones are shown in Table 1.3 (Pachauri, 1992).

An analysis: The geological reserves of a number of known hill deposits and their viability for commercial exploitation have been reported. A review of the geological records indicates that the Himalayas also have similar mineralized belts, which can be exploited commercially. But the reasons given in the following (Mukherjee, 1983) go against the belief that the mineralization in the Himalayas has economic viability. It is stated that the collisional nature of the mountain ranges seems to go against the occurrence of widespread commercial mineralization unless these are inherited by pre-tertiary geological processes. Guild (1971) and Silitoe (1969) have inferred that collision orogens are rather poor in mineral deposits. The structural geometry and extension of the lesser Himalayan mineral occurrences are usually complex, as these are affected by pre-Himalayan tectonic grain as well as the Himalayan fold and thrust (Auden, 1934; Krishnan, 1982). Thus, the strike lengths are invariably low, broken up and suddenly terminate (Saklani, 1978; Valdiya, 1984).

In brief, the Himalayas are a land of great promise from the mineral resources point of view (Anon, 1976). Two schools of thought have emerged out of analysis: (a) First, minerals are found but they are sparsely available and their continuity (large-size deposits/veins/lodes, etc.) is disturbed due to mountain-building forces, tectonic activities and several other geological

TABLE 1.2

Economically Exploitable Minerals of the Himalayas for Commercial Purposes

Mineral Name	Gold	Coal	Limestone, Clay and Kaolin Deposits	Sulphide Ore Deposits	Soapstone	Porphyritic Copper	Glass Sand	Rock Phosphate Deposits
Location	River sands of Assam and Arunachal Pradesh (Subansiri river sands)	Tertiary deposits of Eocene age Northeastern Himalaya, Assam (Anon, 1973)	Limestone in all parts of the Himalaya Mikir Hills in Assam for clay and kaolin (Anon, 1973)	Askot, Pithoragarh district, Uttarakhand	Almora, Uttarakhand (Kumaon Himalaya)	Carbonate rock beds of Krols, Shali, Deoban and Jammu, limestone beds (massive deposits) Uttarakhand	Mikir Hills in Assam, northeastern Himalaya	Garhwal Himalaya, Uttarakhand

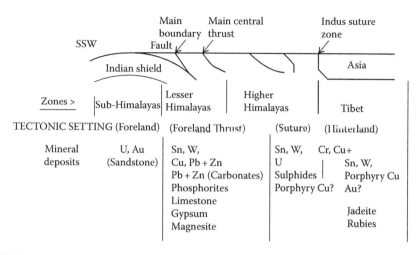

FIGURE 1.6

The Himalayan zones, tectonic settings and mineral deposits. (After Pachauri, A.K., *Himalayan Orogen and Global Tectonics*, A.K. Sinha (ed.), Routledge, Oxford, U.K., pp. 267–288, 1992.)

processes that are still going on in the Himalayas. This is the reason why they are proving uneconomical from the exploitation angle. (b) Second, the abundance of metallic minerals is limited. Thus, it is certain that the Himalayas as a whole are not highly mineralized; however, some minerals are in abundance, for example limestone. Detailed prospecting and exploration are still lacking to prove the mineral reserve quantities even for the number of identified deposits of the Himalayas because of the difficult field conditions. More detailed analysis about the assessment from a commercial angle is given in the next section.

1.5 Assessment of Commercial Exploitation

To get further insight into the viability of various minerals from the exploitation angle, the following points should be noted. The mineral potential and overview of the mineralization in different locations of the Himalayas, as described earlier, can be supportive for the purpose of overall assessment.

1. The Indian Himalayas contain a variety of mineral resources and are a storehouse of some minerals in large quantities. But the number of known mineral deposits of the Indian Himalayas has not been examined in detail from the commercial exploitation angle.

2. The Himalayan mineral deposits are complex in structure and situated in difficult and dangerous locations, making their exploitation a tough and stupendous task.

TABLE 1.3

Ore/Minerals in Different Tectonic Zones of the Himalayas

S. No.	Tectonic Settings (North to South)	Ore/Minerals	Remarks
1.	Precollision basins (between the MBT and the MCT)	No economically valuable minerals have been found, but Au is possible.	This zone is difficult to distinguish in the field.
2.	Ophiolitic suite of rocks	Chromium and chromite; Cu-bearing minerals and Cu–sulphide minerals.	Mineralization is expected to be associated with mafic rocks.
3.	Hinterland region (north of ITSZ)	Rubies, sillimanite and gneisses; porphyry copper, tin–tungsten and gold deposits.	High-grade metamorphosed zone, for example Kashmir. Huge batholiths are found.
4.	Barren hinterland basins	No mineral deposits.	Uranium deposits can be expected.
5.	Foreland thrust belt zone	Tin–tungsten, sulphide deposits, Pb and Cu, chalcopyrite, pyrite, arsenopyrite, pyrrhotite and galena.	Collision-related granites host the most amount of ore mineralization.
6.	Foreland basin	Uranium, vanadium and copper; uranium in the form of uraninite in Shiwalik region sandstones and greywackes.	'Placer gold' can be found in the Himalayan streams with its origin in the higher Himalaya.
7.	Sedimentary sequences in the Klippe zone	Limestone, coal and magnesite. Carbonate deposits host some metal mineralization, such as lead–zinc.	Carbonate deposits occurring in this zone are not economical from the exploitation angle.

Source: Pachauri, A.K., Plate tectonics and metallogeny in the Himalaya, in: Sinha A.K., ed., *Himalayan Orogen and Global Tectonics*, International Lithosphere Programme Pub. No. 197, 1992, pp. 267–288.

3. Limestone is the most abundantly found mineral of the Himalayan region (1184 million tons of proved reserve; UNFC 221 and 331 category; Soni, 1997) and is available in nearly all states of India over which the Himalayas extend.

4. The Himalayan region as a whole is a tectonically active region with a number of geological discontinuities, that is faults, folds and thrusts. The strike length of mineral deposits is invariably low and broken up and suddenly terminates. Therefore, big deposits are divided into small patches and found in the form of small deposits.

5. Commercial exploitation of low-grade ores has not been attempted so far. With the state-of-the-art technology of mineral winning and mineral benefaction (modern technologies), commercial solutions for exploitation of such ores and minerals can be easily achieved. These new initiatives/attempts can become resource earners for the state in which these minerals lie.

6. The Himalayas are the cordilleras of mountains with young and weak rock formations. From the ecological and environmental angle, they are sensitive to human disturbances. Overexploitation or change in the natural pattern of biotic and abiotic environment has far-reaching consequences on the plains down below. This has imposed restrictions for their commercial exploitation and made this topic debatable and controversial.

7. Exploitation of mineral resources in EFAs, which are rich in genetic resources, obviously results in a series of chain effects such as soil erosion, floods in the plains, loss of rare and endangered species of both vegetation and wildlife, loss of biomass and destruction of scenic beauty. Their exploitation beyond a limit warrants attention and restriction on their commercial excavation.

8. Mineral winning activity requires a huge land area. In EFAs, land is restricted and open space is short because of topographical reasons. Thus, diversified and productive output of the land in the form of mineral exploitation takes the back seat.

9. For the hill population, meeting the demand of fuel and fodder is the top priority, and this makes agriculture an extremely important land use issue. This also restricts the land resource use for mineral extraction, as it is a capital-intensive activity.

An analysis from the industrial perspective, as was done in the previous section and by others (Anon, 1973, 1976), indicates two schools of thought that contradict each other. But both schools have ascertained that mineralization exists prominently in the Himalayas, and on case-to-case basis, it should be carefully evaluated for commercial exploitability.

On the basis of the analysis and discussions given earlier, it can be easily concluded that the Himalayan region contains mineral resources of commercial importance. Both large-scale economic mineral deposits and mineral deposits of trace elements are available in the Indian Himalayas. Such available deposits need to be first thoroughly analyzed and evaluated scientifically (on case-to-case basis) for the purpose of their commercial exploitation. Without proper evaluation and cost–benefit analysis, the minerals of the Himalayas should remain as reserve only. These are the treasure troves of the country in which they are vested. Their selective exploitation will save the serene hill environment, which is more precious than the profits that the hill people will get through their exploitation.

1.6 Mining in the Himalayas

The Himalayan chain is spread across a number of Asian countries and is no doubt rich in minerals, and the possibilities of finding more minerals are immense. In India, systematic efforts to locate mineralized zones in a proper manner started only after 1947. There were evidences of earlier mining efforts at some places in the Himalayas, where mineral reserves in significant quantity (at later dates) were proven scientifically.

In the Himalayas, mining activity is being carried out mostly on a limited scale by private entrepreneurs. Thus, small-scale mining dominates in the whole of the Himalayan region. Some registered government companies and big private cement companies (e.g. Almora Magnesite Limited, Ambuja Cements, Jaypee Cements) operate their mines in the Indian Himalayas. A review of the mining practices in the hilly areas of Nepal (DMG, 1991), Bhutan (Shapkota, 1991), China, Pakistan and India has indicated that the Himalayan mining is mostly surface mining starting from the hilltop to the bottom, that is 'open-cast stripping'. 'Hill mining on slopes' and 'open-pit mines' are also reported from this region. Some noted underground mines (e.g. the base metal mines in Sikkim Himalayas and rock phosphate mines [Durmala and Maldeota mines] near Dehradun, UK [Indian Himalayas]) were operative previously but closed in 2015.

A mineral-wise list of mines of the Indian Himalayas operative in various districts and their status of operation are given in Table 1.4.

Since the mines of the Himalayan region are largely owned by private owners in the unorganized sector, their operational status keeps on changing with time quite frequently. Some prominent mineral-bearing areas of the Himalayas and their locations are given in Figure 1.7. These Himalayan deposits are of economical importance and are viable from the mining angle (see Table 1.1).

TABLE 1.4

Mineral/District-Wise List of Mines of the Indian Himalayas and Their Status of Operation

S.No.	Mineral	State/District	No. of Mines	Status		Type of Mines			
				I	II	A	B	C	D
1.	Barytes	HP/Sirmour	3	–	3	–	–	3	–
2.	Bauxites	J&K/Udhampur	2	–	2	–	–	2	–
3.	Multimetal (Cu–Pb–Zn)	Sikkim Himalaya	2	–	2	–	2	–	–
4.	Limestone	J&K/Udhampur	1	–	1	–	–	1	–
5.	Clay	J&K/Pulwama	1	–	1	–	–	1	–
6.	Dolomites	Uttarakhand/Tehri Garhwal	12	2	10	–	–	2	–
7.	Gypsum	J&K/Doda	1	1	–	–	–	1	–
8.	Gypsum	Uttarakhand/Tehri Garhwal	1	1	–	–	–	1	–
9.	Limestone	HP/Sirmour	49	15	34	1	–	48	–
10.	Limestone	HP/Himalaya (Sirmour, Bilaspur, Solan and other areas)	67	11	56	04	–	07	–
11.	Limestone	Uttarakhand/Dehradun	2	2	–	1	–	1	–
12.	Limestone	Uttarakhand/Tehri Garhwal	2	2	–	–	–	2	–
13.	Magnesite	Uttarakhand/Almora	2	2	–	2	–	–	–
14.	Magnesite	Uttarakhand/Pithoragarh	1	1	–	1	–	–	–
15.	Phosphorite	Uttarakhand/Dehradun	2	–	2	2	–	–	–
16.	Soapstone	Uttarakhand/Almora	5	4	1	–	–	5	–
17.	Soapstone	Uttarakhand/Pithoragarh	–	4	3	1	–	–	4
18.	Limestone	J&K/Pulwama	2	2	–	1	–	1	–
19.	Rock salt	HP/Mandi	3	2	1	–	1	–	2

Sources: Indian Bureau of Mines (IBM), Dehradun, Uttarakhand, India; Department of Industries, Shimla, Himachal Pradesh, India.

Notes: A, open-cast mechanized mine; B, open-cast mine partly underground; C, open-cast manual mine; D, underground manual mine. I and II mean working mines and closed mines, respectively.

Abbreviations: HP, Himachal Pradesh; J&K, Jammu and Kashmir (States of India).

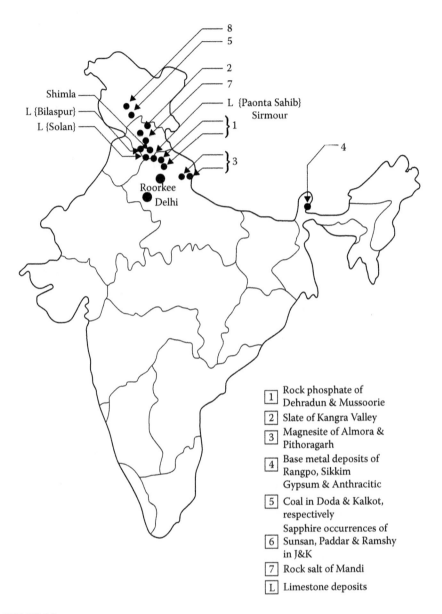

FIGURE 1.7
Mining areas and deposits of economic importance of the Himalayas.

The analysis of geological reserves of various deposits, data of mines and literature indicates that the exploitation of limestone by open-cast methods on small and medium scales is a well-developed practice in the entire Himalayas wherever this mineral occurs for commercial exploitation (both as major and as minor minerals). Conventional mining methods, that is drilling and blasting, which need a large labour force (manual methods) are in vogue. Industry as a whole has realized the typical characteristics of the Himalayan region (as described in this chapter) and the 'protection of the environment at mines' is no doubt an important issue that is well understood by all mine owners and companies irrespective of their size and scale of operation. Even then, financial constraints have posed difficulty in reaching unanimity about finding a midway course that may take care of regional needs, environmental protection and societal goals.

In linking mining in the Indian Himalayas with that in Nepal, Bhutan or China, a mention of limestone mining of the Sirmour district in HP (India) can be made, which is an example of 'open-cast stripping in the Himalayas' executed/implemented by private entrepreneurs. It is an example of small-scale manual/semi-mechanized mining. It may be noted that Sirmour, the southern district of the Indian state of HP (30°22′30″–31°01′20″ N and 77°01′12″–77°49′40″ E), came into limelight after stoppage of limestone mining in the Doon Valley in early 1990s and became the prominent mining industrial belt for the surrounding states of UP, Haryana, Punjab and HP (Solan and Shimla), which need limestone as the raw material. The district covers an area of 2825 km^2 in the lesser Himalayas and Siwalik ranges and shows a rugged mountainous terrain with moderate relief. Three rivers, namely the Giri, the Tons and the Yamuna, control the drainage pattern of the region and the district that encompasses the mining areas as well.

The mining operation in the Sirmour district can be divided into seven zones. The limestone of various grades exists both in trans-Giri and in cis-Giri areas of Sirmour district. In the trans-Giri area, its occurrence extends throughout the northwest–southeast length of the district and is well exposed at various places, that is in the vicinity of *Nohra, Sangrah, Kamraoo* and *Shiva-Banor*, whereas in cis-Giri, the limestone for industrial uses is confined only to a limited area, that is in the central and southeastern portion of the district. There are 52 registered/listed small-scale mines (lease granted by the government) in Sirmour district alone, out of which 49 are of limestone and 3 are of baryte. Among these 52, presently, 15 limestone mines are operational (Table 1.4). The limestone produced from these mines is used for the kilns making lime as the raw material feed. In 1999–2000, more than 67 mines of limestone alone were operative in the Sirmour region. In 2014–2015, this number was reduced because a number of them had not obtained proper permission from the state authorities. The Cement Corporation of India's Rajban Cement Plant's captive mine (Manal Limestone Mine) is the biggest mine of the Sirmour area producing 1,00,000 tons of limestone per year. Total limestone production from the Sirmour district of the HP Himalayas is 9,04,732 tons per annum, which keeps fluctuating depending on the market demand. The product of these mines is treated as minor minerals.

The major observations in manual mines of the Himalayan region are as follows:

- No detailed exploration was done prior to the start of mining operation.
- Face height varies from 3 to 18 m, and due to high faces, formations of overhangs are observed; openings of working faces on overhanging slopes are also noticed at many places.
- The faces are often irregular and not smooth.
- Sometimes, limestone is extracted selectively, leaving the lean zone as such.
- Mining plans are not followed in practice in some mines.
- Rock masses are foliated and jointed and are not very strong except in highly calcined patches.
- They lack even minimum machinery required for systematic mining.
- Selective mining along the roadside is carried out, which is dangerous for the safety of both humans and machines, and even causes road blockage.

Various practices adopted by small-scale mines lack proper planning. Age-old methods have resulted in environmental degradation and ecological damage, terming these mines as 'unscientific'. Several lacunae of existing practices are thus identified and described in Chapter 2 (see Sections 1.6 and 1.7).

1.7 Current Status of the Mining Practices in the Indian Himalayas

India is an important cement producing country in the world, and a portion of its total limestone production is from the Indian Himalayas. Because of its ample availability and economic viability, limestone extraction in the Himalayas is profitable for industries. The Himalayan limestone is of very high grade and suitable for paper, sugar, cement and metallurgical industry and is of significant importance. The $CaCO_3$ percentage of the Doon Valley limestone is as high as 99.4% with no deleterious constituents. Adequate demand and sufficient reserve of limestone in the lower Himalayas have made its extraction possible in some concentrated areas which are listed here:

- Limestone belt of Doon Valley near Dehradun, UK
- Limestone belt of Bilaspur, Sirmour and Solan districts, HP

- Limestone belt of Tehri Garhwal area of UK (erstwhile UP)
- Limestone belts of Udhampur and Pulwama districts in J&K
- Limestone belt in the state of Assam (Nawgaon, North Lakhimpur, Jorhat and Dibrugarh) in the northeastern region

From the mining point of view, Indian limestone extraction enterprises can be broadly divided into three categories: large corporate sector, state-owned corporate sector, and small private sector. Four large mines of corporate sector are operative in the region, namely Jaypee Cement's Captive Mine (near Baga Plant, district Solan); Associated Cement Company's (ACC) Gagal Limestone Quarry at Barmana, district Bilaspur, HP; Manal Limestone Quarry of Cement Corporation of India Limited (CCI) near Rajban in Sirmour district, HP and Himachal Ambuja Cement's Kashlog Limestone Mine near Darlaghat in Bilaspur, HP. All these mines make use of their limestone production in cement and clinker production. Earlier records indicate that the National Mineral Development Corporation had proposed to mine Arki deposit in the Solan district of HP in the late 1990s, which was later given to the Jaypee group. Jaypee's Himachal Cement Plant (JHCP) commissioned its Baga–Bhalag Limestone Mine (Figure 1.8a) in 2010 in the Baga Village of Tehsil Arki, district Solan, HP. JHCP has 2.54 million tons cement production capacity annually (Box 1.2).

In Doon Valley, the Lambidhar Mine of the UP State Mineral Development Corporation (UPSMDC) near Mussoorie in the state of UP was another big mine owned by the state government, but this mine was closed in January 1996 on account of environmental reasons and expiry of the mining lease.

Three cement plants, namely JHCP, ACC's Cement Plant (5.5 km NW of JHCP) and Gujarat Ambuja Cement Plant (8 km SE of JHCP), have contributed immensely to the overall industrial growth of the hill state of HP through direct and indirect employment, including ancillary business opportunities related to cement production.

In Tehri Garhwal district, two mines owned by small private entrepreneurs are operative with an annual production of about 90,000 tons. J&K Cements Limited and J&K Minerals, both fully owned government companies, carry out the mining of limestone for their captive cement plants (600 tons per day capacity dry process cement plant) in the Khrew area of Pulwama district of J&K. Its salient features are available at http://www.jkcl.co.in/about.php and http://www.jkminerals.com/about.htm.

The Sirmour district of HP is another prospective area of small private sector mines, particularly for limestone. It may be noted that limestone mining in this area got a boost after the closure of limestone mines in the Doon Valley. In all, seven important and significant limestone-bearing areas exist in Sirmour district where the cluster of mines is situated. Nearly, all of them are in the small private sector except one, which is CCI's mine of the Rajban Cement Plant (Manal Limestone Mine). These limestone-bearing areas are

FIGURE 1.8
(a) Baga–Bhalag Limestone Mine in the Indian Himalaya. (b) Environmental measures taken at the Baga–Bhalag Limestone Mine.

Sataun-Kamraoo, Manal, Pamta-Bohar, Rajpura-Bharli, Sangrah-Bhootmari, Shiva-Rudana and Nohra (Figure 1.10). The mines of these areas are mere quarries and are being worked by conventional open-cast, manual mining methods. Their year-wise production varies from 5,000 to 50,000 tons, and the employment statistics fluctuates widely depending on the work and demand of limestone in the market. Very small mine owners (<5 ha lease area) produce between 50 and 100 tons of limestone per day in these areas.

BOX 1.2   JAYPEE HIMACHAL CEMENT PLANT

Jaypee Himachal Cement Plant (JHCP), shown in Figure 1.9, is the newest cement plant of HP commissioned in 2010. This plant is located in the Baga Village of Tehsil Arki, district Solan, and has a planned production capacity of 2.54 million tons (cement production) annually, which requires limestone of 3.1 million tons per year from its captive mine (Baga–Bhalag Limestone Mine) located at about 0.57 km from plant and at Bhalag village.

BAGA–BHALAG LIMESTONE MINE

The Baga–Bhalag Limestone Mine is a captive mine of Jaypee's Himachal Cement Plant (a unit of Jaiprakash Associates Ltd.). This captive mine can be located on Survey of India Toposheet No. 53A/15 in Baga and Bhalag villages of Tehsil Arki, Solan, HP, and lies at a distance of 42 km from the town Bilaspur and around 67 km from the state capital, Shimla.

Geologically, this deposit belongs to *Sorgharwari*, member of *Shali formation* of the pre-Cambrian age. The deposit contains pink and grey limestone. Shale present in this area occurs as synclinal structure with dolomite of the Khatpole member on either side of the syncline.

FIGURE 1.9
Jaypee Himachal Cement Plant at Bagga, Solan, Himachal Pradesh.

The total area under mining lease is 331.424 ha and mining is being carried out in two blocks, that is Bhalag block on the north side and Baga block on the south side. The mines are spread between the latitudes 31°18′50″ N and 31°20′43″ N and longitudes 76°53′11″ E and 76°54′46″ E. The deposit is of hilly type and limestone occurs on the slope (the entire hill is composed of limestone); hence, the mine is being worked from top to bottom by the 'top slicing method'. This mine has no overburden, and the limestone is fully outcropped. The JHCP mine is a fully mechanized mine and equipped with a fleet of modern machines, for example PC 600 (Komatsu) excavators, articulated dumpers A35E (Volvo), ROC-L8 (Atlas Copco) drilling machines, ROC-F9 (Atlas Copco) drilling machines, Sandvic DX 800 (Sandvik) drilling machines and D275 A (Komatsu) dozer. Blast hole drilling is carried out by down-the-hole drilling machines equipped with dust collectors and water mist dust suppression systems. The unit operation of blasting is being carried out using NONELs to control ground vibrations and fly rocks. The blasted mineral is excavated and loaded by Komatsu PC 600 Backhoe excavators into Volvo A 35E articulated dumpers. Minerals are transported from mine faces and unloaded into the crusher. The production details achieved are as follows.

Annual Production at Baga–Bhalag Mine[a]

S. No.	Year	Limestone	Shale	Total
1.	2012–2013	37,89,261	5,44,065	43,33,326
2.	2013–2014	33,51,924	4,10,549	37,62,473
3.	2014–2015	33,54,162	4,19,129	37,73,291

[a] All values are in tons.

Ground vibration and air blast overpressure are monitored for different blasts as per the scientific norms recommended by the authorized agency (CIMFR). The blast designs are prepared as per the encountered rock/ground conditions.

The Baga–Bhalag Limestone Mine, being a newly opened deposit limited degraded area due to mining, is available for reclamation. However, the reclamation and rehabilitation of the mined area are done as a routine and continuous process in the mine. The excavated area is rehabilitated in a planned manner by adopting large-scale afforestation (Figure 1.8b).

JHCP is the big industrial unit of HP Himalaya and makes use of the latest technology and consumes limestone as the raw material feed.

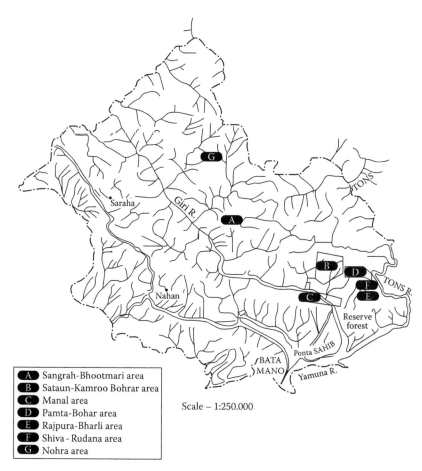

FIGURE 1.10
Map indicating zone of concentrations of mining leases in Sirmour district.

In 2015, there were 15 operative mines of limestone in the Sirmour region. The Manal Limestone Mine, Cement Corporation of India's captive mine, is the biggest mine of the Sirmour area, where mechanized mining by dumper shovel is done. Rest all leases that operate in the area are by private mine owners. All of them carry out mining operation on the small scale.

Small mining entrepreneurs in HP acquire leases from the Department of Industry (Geological Wing), Shimla. Two types of practices are adopted for the grant of mineral rights to private owners. One is the 'mining lease grant' for excavation of in-situ hill mass consisting of limestone and the other one is the 'grant of permit' for lifting of already excavated limestone that is lying at various places or generated during road construction (such practices are permitted as these are considered good from the mineral conservation point of view). Manual open-cast mining methods are the general practice. The overburden is removed by manual means using low explosives, which are locally available, to form small-size benches. The exposed mineral body is exploited by manual cutting using picks and crowbars and transported by trucks and dumpers. The bench height varies from mine to mine, and no specific height is maintained. Drilling is done by jackhammer drills only.

The mining of limestone in Sikkim and Arunachal Pradesh and in the Darjeeling Himalayas has been scantily reported. Quarrying through minor mineral lease grant to private parties and to captive cement plants (both medium- and small-size mines) is a normal practice.

As described earlier, limestone mining areas are concentrated in various districts of UK, HP and J&K and the NE region, and a number of mines are invariably owned by small private entrepreneurs. The production, employment statistics and other details for this sector are not described here because this sector is highly vulnerable to frequent changes. For example, in the late 1970s, the Doon Valley had more than 105 small mining leases, whereas in the 1980s, on account of public interest litigations, nearly all the mines were closed and only one small mine, Ladwalkot, of M/s Rakesh Oberoi was permitted to continue up to 1995 (Appendix 1.A).

When the mining practices in the other part of the Himalayas, that is Sikkim Himalaya (NE Himalaya), are reviewed, it is found that the most prominent base metal deposits of the Bhotang multimetal mine (near Rangpo) and Dikchu Cu–Zn mining project of Sikkim which were under exploitation by the Sikkim Mining Corporation (SMC) till 2006 are closed now (Box 1.3). Similar condition about the underground rock phosphate mines near Mussoorie in the Garhwal Himalayas is observed.

Dolomites, talc/soapstone, barites, gypsum, quartzite, etc., are the other economic minerals available in the Sikkim Himalayas. It clearly shows that it is not the mineral and its availability in the Himalayas that are in question but its eco-friendly and scientific exploitation that is the major concern.

Periodically, the Indian regulatory agencies have observed that the deteriorating environmental conditions in the Himalayan mining areas are aggravated because of mine owners flouting the existing laws for economical gains. This causes a negative impact and calls for the federal/state government intervention. Because of these reasons, mining activities had to be banned in many mines, leading to the stoppage of total mining operation. In sensitive and fragile areas of the Himalayas, restricting/banning mining for various types of minor and major minerals, stone quarrying and crushing except for the domestic needs of bona fide local residents was found to be helpful in improving the ecological and environmental condition. (Here, 'bona fide local residents' means someone who is residing together with his or her minor children in that area for an uninterrupted period and who is on the electoral roll on the date of the government notification under force or

BOX 1.3   POLYMETAL ORE DEPOSIT
OF THE SIKKIM HIMALAYAS

In the Sikkim Himalayas, which are in the northeastern part of the Indian Himalayas, the base metal deposits of Cu–Pb–Zn at Bhotang, Rangpo, were under exploitation till 2006 by the SMC. The Dikchu (Cu–Zn) project (ore body length × vertical depth × thickness = 564 m × 100 m × 1.25 m; 90%–93% recovery of Cu in the concentrate; ore reserve = 0.45 million tons with 2.82%–3.0% Cu and 0.90%–1.0% Zn) was another polymetal deposit under consideration for exploitation by the SMC. Both these deposits are economically viable and suitable for commercial exploitation.

The Bhotang mine is located near Rangpo in the Sikkim state of India and worked by underground mining methods using stoping. The mine (east Sikkim) was started in 1966–1967 and was the only multimetal mine of the Indian Himalayas producing three metals: copper, lead and zinc from one complex ore (2.82% Cu and 0.90% Zn). The Bhotang ore also contains appreciable amounts of silver and some amount of gold.

The mine was owned by the SMC, a joint venture of the government of Sikkim and the government of India. The recoverable reserve as on 1 April 1997 was reported to be 2.44 lakh tons. With a work force of 50–100 persons (approximately), the mine was achieving a total production of 45–100 tpd (tons per day) only. Records show that in this underground mine, extracted stopes were located at the third and fourth levels. The mine development work was carried out up to the fifth level and below to extract ore deposit below the river Tista. The mine management was using underground mine supports of concrete and steel props inside mine instead of timber, and the mineral processing plant waste was being used for back filling the mines.

The Bhotang mine operation was earlier confined to the hill section, but the plunge of the ore body led to active mining operation below the river Tista, leaving a safe pillar barrier of about 60 m in between the mine workings and the river. In view of the earlier discussion, the mine was equipped with geotechnical instruments, such as linear variable differential transformers (LVDTs), extensometers, lead cells and vibrating wire stress meters, and being worked below the river, under technical guidance.

The Cu–Pb–Zn concentrate produced after the treatment of the ores and beneficiation at the Rangpo plant was sold to HCL, Ghatsila, and HZL, Visakhapatnam, for further processing. Attempts were made to obtain by-products such as lead chloride and lead chromite, with the help of Hindustan Copper Limited and other private parties through R&D. Since this mine was located in the Himalayas, additional care was taken for land and environment protection by SMC. The degraded land area on surface was revegetated using plantations, the work on which was done till 2014 by the Sikkim government and mine management.

The reasons for closure of the mine were many: (a) the high bismuth content in lead concentrates and the presence of other impurities that has no potential market; (b) the difficult mining condition in stopes and safety; (c) very low volume of production and (d) other managerial reasons.

Sources: ENVIS, Sikkim's website; IBM, *Indian Minerals Yearbook-2012, Part I: Sikkim State Reviews, 51st ed.,* Indian Bureau of Mines (Government of India), Ministry of Mines, Nagpur, India, 2014; Communication with SMC (May, 2015).

consideration.) However, for socioeconomic reasons and on a limited scale, such activities (mining, stone quarrying and crushing) should be permitted based on site-specific evaluation. Some special areas, for example very steep hill slopes (>70°) or areas with a high degree of erosion, spring lines and groundwater recharge, are strictly not allowed for development intervention by either the federal or the state government.

In brief, recent assessment indicates that a number of the Himalayan mineral deposits that were exploited for commercial purpose in the late 1990s were non-operative in 2015, for example Maldeota and Durmala Rock Phosphate Mines, Mussoorie; Lambi Dehar Limestone Mine, Mussoorie; Bhotang Mine, Rangpo, east Sikkim and Coal of Doda and Kalkot (J&K). It was observed (from the mining point of view) that complex geological situation, weak rock formations, discontinuous ore bodies, unproved ore reserve quantities and patchy mineral pockets that were uneconomical have led to the closure of rock phosphate underground mines of Dehradun and Mussoorie. Similar situation with the base metal deposits of Rangpo, Sikkim, was observed. Review and analysis of the mining activities and its current status (described in the previous two sections) thus clearly show the following:

1. Large-scale mining in the Himalayas is a planned activity, but small-scale mining (artisanal mining) sector comprising small entrepreneurs is largely unplanned and unscientific. This small-scale mining, which was previously considered to be wasteful mining due to its unsafe and unhealthy condition, brutalizing labour practice, and damage to the ecology and environment, can be turned into a safe and scientific practice (MMSD, 2001).

2. With reference to all those developing countries that lie in the Himalayas, such scientific practices for small-scale mining (refer Section 6.2) can play a major role for the country's economic development because of their small investment, manageable environmental conditions and more direct employment.

3. It is almost certain that when high-grade ores of plain areas are mined out to the maximum possible extent, or exhausted, low-grade and geologically complex deposits of hilly Himalayan terrain can be harnessed. Probably they will meet the future mineral needs.

1.8 Legal Repercussions of Mining in the Himalayas

As mentioned earlier, limestone is the most exploited mineral in the Indian Himalayas. The mining of limestone in the Himalayas had legal repercussions as well. In this subsection, such cases are briefly analyzed, and the significant among them are as follows:

- The Supreme Court case of Doon Valley
- The Saproon Valley (Solan district, HP) case of HP High Court
- The Sirmour district legal case of the HP High Court

The first two cases resulted in a 'ban on mining', whereas the third culminated in a 'ban on blasting' in limestone mines and the protection of the environment from mining activities.

The Doon Valley was in limelight during 1983–1985 on account of a public interest litigation in the Supreme Court of India for issues relating to environment and ecological balance due to limestone mining. This case has brought into sharp focus the conflict among development, environmental protection and conservation. This was the first case of its kind in the country, and it opened new vistas of environmental protection in the Indian Himalayas and became a case of significance not only to the people residing in the nearby areas but to the whole country and the people of India. The case resulted in the closure of nearly 108 mines (both big and small) in the Doon Valley. Only two mines, the Lambi Dehar Mine and the Ladwalkot Mine, were permitted to operate up to 1995 and 1994, respectively.

In 1988, the Saproon Valley in the Solan district of HP was dragged into a legal battle by the local public in the High Court of HP. The court observed the following:

- The limestone found in the area is not of very high grade (lime kiln grade), and only eight mines were affected.
- Direct employment benefits were available to about 250 families.

- Royalties from these eight mines amounted to the tune of Rs. 1,00,000 in 1988, which rose to about Rs. 4,00,000 on 1 April 1991, as a result of increase in the rates of royalty.
- Mining at one mine was carried out along the hill slope on the Solan–Kalka highway and was an eyesore.
- For the transportation of nearly 30,000 tons of limestone, the mine owners contributed nearly Rs. 4,50,000 per annum in the form of tax to the exchequer.
- The direct and indirect pecuniary advantage to the state from the mining operation in all these eight mines was quite low.

The court compared Doon Valley case with the Saproon Valley case and observed the following:

1. In the earlier case of Supreme Court on the Doon Valley, nearly 108 leases were being affected, whereas in the Saproon Valley case, only eight mines were under consideration.
2. The mining operations of the Saproon Valley started only 10 years earlier in 1978, whereas in the Doon Valley area it was nearly 75 years old (started in 1911).
3. In Saproon Valley, no mine was located within any reserve forest area and the mines were out of municipal limits (at a distance of 3 km from municipal limits).

Keeping these points in view, the court directed for the closure of the eight mines in the Saproon Valley. However, some mines were directed to comply with the stipulations of the high-powered committee. The court appointed a 'monitoring committee' to keep a watch and report periodically.

Another significant social litigation in the High Court of HP was that in 1987 by a local lady Kinkri Devi of Sangrah village, district Sirmour, citing the ill effects of mining with the following charges:

1. Limestone mining operations in various quarries in Sirmour district, particularly in and around Sangarh village, in Tehsil Shillai and Paonta Sahib, were causing havoc to the adjoining land, water resources, pastures, forests, wildlife, ecology and inhabitants of the area.
2. The blasting operation carried out by the lessees was causing damage to the village houses and civil structures in the nearby areas.
3. Due to the blasting operation, groundwater table was going down, leading to the drying up of wells, springs and other water sources.

On 26 November 1987, the deputy commissioner of Sirmour issued a communication on the directive of HP court order dated 3 November 1987 to the mining officer of Sirmour to ensure the stoppage of blasting operation in all the mines operating in the district.

On 16 August 1991, this case was finally decided in favour of the mine owners by lifting the ban on blasting and the appointment of a high-powered monitoring committee. The court ordered that the committee must visit the mines of Sirmour area within 2 months and enquire whether the provisions contained in the approved mining plans were being faithfully followed by the lessees concerned. Thereafter, at the intervention of the High Court, the MOEF, government of India, ordered detailed scientific investigations of the limestone mining areas and worked out the impact of limestone mining on the local population, including the preparation of an Environmental Management Plan for the Sirmour district as a whole. The author was involved in this work on behalf of his institution and played a significant role.

Note: For the preparation of this section, the following legal documents were referred. Since these are categorized as 'classified documents', they are not listed in the reference list and given here separately. All these documents are in the public domain.

- *Civil Writ Petition, Copy of the Judgment of the HP High Court and the Supreme Court of India.*
- *Supreme Court Petition Nos. 8209 and 8821 of 1983 decided 12/3/1985. Petition Nos. 16437, 16683, 16757, 17895, 17896 and 18244 of 1985 decided on 13/5/1985. (Rural Litigation and Entitlement Kendra, Dehradun and others versus the State of UP and others, AIR 1985, SC1259 and 652.)*
- *HP: CWP No. 82 of 1987, Kinkri Devi and High Court of Shimla. Others versus the state of HP, decided on 16/8/1991.*
- *CWP No. 567 of 1988 (with CWP Nos. 598, 599, 600 and 601). General Public of Saproon Valley and others versus the state of HP and others, decided on 24/4/91.*
- *Joshi, Chetan (1991), Limestone Mining Operations and Its Implications on Degradation of Environment, report submitted to HP High Court, Shimla.*
- *Notification No. S.O. 923(E) dated 6 October 1988, Published in the Gazette of India, extra part II section, 3(ii), pp. 5–8.*

Thus, it is evident that the Indian Himalayas have live mining practices on small as well as large scale. The region has mineral potential, too, for commercial exploitation. Since the minerals of the Himalayas are the national concern (including for all those countries in which the Himalayas are extended), they are to be harnessed with precaution. Further, mining has both positive and negative environmental impacts. In order to satisfy the mineral needs of the society and balance the ill effect of mining, the sustainable development concept and integrated solutions must be adopted in practice.

The Himalayan region(s) have a rural-based economy; hence, to plan and manage the hill population and hilly region, scientific practices are essential. The mining and mineral-based industrial activities can significantly improve the resource development of the region in future.

In a nutshell, mining is essentially an economic development activity undertaken for the overall good of the people; therefore, mining should coexist with the society and carried out in such an eco-friendly fashion that it is acceptable to the maximum population. A balance between the losses and gains of resource exploitation must be achieved. Affected land mass, water, air and the ecology as a whole should be protected, and the community should adopt the attitude to absorb losses, if any.

Appendix 1.A: Mining in Doon Valley

Review of old published record indicates that the famous Doon Valley of the lower Himalayas/Siwaliks was once a prominent place in the Himalayas from a mining view point. The Doon Valley is bounded by the Himalayan Frontal Thrust to its south and the Main Boundary Thrust to its north. Several mineral exposures/outcrops in the field about the lesser and higher Himalayan sequences makes Doon Valley a prominent mineral resource region.

Mining in the Doon Valley started on a small scale in 1936 and gained momentum after 1947. Initially, mining commenced on the basis of the permits acquired from erstwhile UP government (now Uttarakhand) and, subsequently, since 1962 on the basis of the mining lease granted under the Mineral Concession Rules 1960 and the UP Minor Mineral Concession Rules 1963. In course of time, more than 105 mining leases came into existence. Since the mining areas were mainly in the vicinity of tourist attraction points of Dehradun and Mussoorie, this limestone belt of the Himalayas came into the limelight and faced the wrath of the public on account of environmental degradation. Local residents, particularly in the vicinity of mining areas, raised objections in 1974 against the ill effects of mining. But due to the lack of statutory provisions, nothing much could be done to mitigate the grievances. In 1982, when most of the leases were to be renewed, a strong objection against the continuance of mining in this environmentally fragile area and tourist hill resort was raised by the local residents and social organizations. Besides representations to the government organizations, a public interest litigation was also filed in the Supreme Court of India in 1982–1983. As a result of the verdict of the court, most of the limestone mines were closed. In this context, the Doon Valley notification (Appendix 1.B) issued by the MOEF, Government of

India S.O. 102(E), No. J-20012/38/86-IA dated 1 February 1989, is extremely significant. As per this notification, guidelines were issued for permitting/ restricting industrial units in the Doon Valley area and industries were classified based on their environmental and ecological impacts (www. moef.nic.in/legis/eia). As per this legislative order, prior approval of the Union Ministry of Environment and Forests before starting any mining activity in Doon Valley is required.

Appendix 1.B: Doon Valley Notification Dated 1 February 1989

(*Source:* www.moef.nic.in/legis/eia)

Notification under 3(2)(v) of Environment (Protection) Act, 1986, and Rule 5(3)(d) of Environment (Protection) Rules, 1986, restricting location of industries, mining operations and other development activities in the Doon Valley in Uttar Pradesh.

S.O. 102(E) – Whereas notification under subrule (3) of rule (5) of the Environment (protection) Rules, 1986, inviting objections against the imposition of restriction on location of industries, mining operations and other developmental activities in the Doon Valley, in Uttar Pradesh was published vide No. S.O. 923(E), dated 6 October 1988.

And whereas all objections received have been duly considered by the Central Government:

Now, therefore, in exercise of the Powers conferred by Clause (d) of sub-rule (3) of Rule (5) of the said rules, the Central Government hereby imposes restrictions on the following activities in the Doon Valley, bounded on the North by Mussoorie ridge, in the North-East by Lesser Himalayan ranges, on the South-West by Shivalik ranges, river Ganga in the South-East and river Yamuna in the North-West, except those activities which are permitted by the Central Government after examining the environmental impacts:

(i) Location/siting of industrial units – It has to be as per guidelines given in the annexure or guidelines as may be issued from time to time by the Ministry of Environment & Forests, Government of India.

(ii) Mining – Approval of the Union Ministry of Environment & Forests must be obtained before starting any mining activity.

(iii) Tourism – It should be as per Tourism Development Plan (TDP), to be prepared by the State Department of Tourism and duly approved by the Union Ministry of Environment & Forest.

(iv) Grazing – As per the plan to be prepared by the State Government and duly approved by the Union Ministry of Environment & Forests.

(v) Land Use – As per Master Plan of development and Land Use Plan of the entire area, to be prepared by the State Government and approved by the Union Ministry of Environment & Forests.

(No. J-20012/38/86-IA)
K.P. Geethakrishnan
Secretary

1.B.1 Guidelines for Permitting/Restricting Industrial Units in the Doon Valley Area

Industries will be classified under green, orange and red categories, as shown here for purposes of permitting/restricting such industrial units in the Doon Valley from the environmental and ecological considerations:

1.B.1.1 Category Green

A. The following is the list of industries in approved industrial areas, which may be directly considered for the issue of no objection certificate without referring to the MOEF) (in case of doubts, reference will be made to the MOEF).

  1. All such non-obnoxious and non-hazardous industries employing up to 100 persons. The obnoxious and hazardous industries are those using inflammable, explosive, corrosive or toxic substances.

  2. All such industries that do not discharge industrial effluents of a polluting nature and that do not undertake any of the following processes:

  Electroplating

  Galvanizing

  Bleaching

  Degreasing

  Phosphating

  Dyeing

  Pickling and tanning

  Polishing

  Cooking of fibres and digesting

  Desizing of fabric

  Unhairing, soaking, deliming and bating of hides

Washing of fabric

Trimming, pulling, juicing and blanching of fruits and vegetables

Washing of equipment and regular floor washing and using considerable cooling water

Separating milk, buttermilk and whey

Stopping and processing of grain

Performing distillation of alcohol, stillage and evaporation

Slaughtering of animals, rendering of bones, washing of meat, juicing of sugar cane, extracting of sugar, filtrating, centrifugating and distillating

Pulping and fermenting of coffee beans

Processing of fish

Filter backwashing in DM plants exceeding 20 kL/day capacity

Pulp making, pulp processing and papermaking

Coking of coal and washing of blast furnace flue gases

Stripping of oxides

Washing of used sand by hydraulic discharge

Washing of latex

Extracting solvent

3. All such industries that do not use fuel in their manufacturing process or in any subsidiary process and that do not emit fugitive emissions of a diffused nature.

Industries not satisfying any one of the three criteria are recommended to be referred to the MOEF.

The following industries appear to fall in the non-hazardous, non-obnoxious and non-polluting category, subject to fulfillment of the three conditions mentioned earlier:

1. Atta-chakkies
2. Rice Millers
3. Iceboxes
4. Dal mills
5. Groundnut decorticating (dry)
6. Chilling
7. Tailoring and garment making
8. Apparel making
9. Cotton and woollen knitwear
10. Handloom weaving

11. Shoelace manufacturing
12. Gold and silver thread and sari work
13. Gold and silver smithy
14. Leather footwear and leather products excluding tanning and hide processing
15. Mirror from sheet glass and photo frame
16. Musical instruments
17. Sports goods
18. Bamboo and cane products (only dry operations)
19. Cardboard and paper products (excluding paper and pulp manufacture)
20. Insulation and other coated papers (excluding paper and pulp manufacture)
21. Scientific and mathematical instruments
22. Furniture (wooden and steel)
23. Assembly of domestic electrical appliances
24. Radio assembling
25. Fountain pens
26. Polythene, plastic and PVC goods through extrusion/molding
27. Surgical gauges and bandages
28. Railway sleepers (only concrete)
29. Cotton spinning and weaving
30. Rope (cotton and plastic)
31. Carpet weaving
32. Assembly of air coolers
33. Wires and pipes of extruded shapes from metals
34. Automobile servicing and repair stations
35. Assembly of bicycles, baby carriages and other small non-motorized vehicles
36. Electronics equipment (assembly)
37. Toys
38. Candles
39. Carpentry – excluding saw mill
40. Cold storages (small scale)
41. Restaurants
42. Oil ginning/expelling (no hydrogenation and no refining)
43. Ice cream

44. Mineralized water
45. Jobbing and machining
46. Manufacture of steel trunks and suit cases
47. Paper pins and U-clips
48. Block making for printing
49. Optical frames

1.B.1.2 Category Orange

B. Here is the list of industries that can be permitted in the Doon Valley with proper environmental control arrangement:

1. All such industries that discharge some liquid effluents (below 500 kL/day) which can be controlled with suitable proven technology
2. All such industries in which the daily consumption of coal/fuel is less than 24 tpd and the particulars emissions from which can be controlled with suitable proven technology
3. All such industries employing not more than 500 persons

The following industries with adoption of proven pollution control technology subject to fulfilling the three conditions mentioned earlier fall under this category:

1. Lime – pending decision on proven pollution control device and Supreme Court's decision on quarrying
2. Ceramics
3. Sanitary ware
4. Tyres and tubes
5. Refuse incineration (controlled)
6. Flour mills
7. Vegetable oils including solvent extracted oils
8. Soap without steam boiling process and synthetic detergent formulation
9. Steam-generating plants
10. Office and household equipment and appliances involving the use of fossil fuel combustion
11. Machineries and machine tools and equipment
12. Industrial gases (only nitrogen, oxygen and CO_2)

13. Miscellaneous glasswares without involving the use of fossil fuel combustion
14. Optical glass
15. Laboratory ware
16. Petroleum storage and transfer facilities
17. Surgical and medical products including prophylactics and latex products
18. Footwear (rubber)
19. Bakery products, biscuits and confectioners
20. Instant tea/coffee and coffee processing
21. Malted food
22. Power-driven pumps, compressors, refrigeration units, fire-fighting equipment, etc.
23. Wire drawing (cold process) and bailing straps
24. Steel furniture, fasteners, etc.
25. Plastic-processed goods
26. Medical and surgical instruments
27. Acetylene (synthetic)
28. Glue and gelatin
29. Potassium permanganate
30. Metallic sodium
31. Photographic films, papers and photographic chemicals
32. Surface coating
33. Fragrances, flavours and food additives
34. Plant nutrients (only manure)
35. Aerated water/soft drink

Note:

a. Industries falling within the identified list mentioned earlier will be assessed by the state pollution control board and referred to the Union Department of Environment for consideration, before according a no objection certificate.

b. The total number of fuel-burning industries that should be permitted in the valley will be limited by 8 tons per day of sulphur dioxide from all sources. (This corresponds to 400 tons per day coal with 1 sulphur.)

c. Siting of industrial areas should be based on sound criteria.

1.B.1.3 Category Red

C. The following is the list of industries which cannot be permitted in the Doon Valley:

1. All those industries which discharge effluents of a polluting nature at the rate of more than 500 kL/day and for which the natural course for sufficient dilution is not available and effluents from which cannot be controlled with suitable technology
2. All such industries employing more than 500 persons/day
3. All such industries in which the daily consumption of coal/fuel is more than 24 MT/day

The following industries appear to fall under this category covered by all the points as mentioned earlier:

1. Ferrous and non-ferrous metal extraction, refining, casting, forging, alloy making processing, etc.
2. Dry coal processing/mineral processing industries like ore sintering beneficiation and pelletization
3. Phosphate rock processing plants
4. Cement plants with horizontal rotary kilns
5. Glass and glass products involving the use of coal
6. Petroleum refinery
7. Petrochemical industries
8. Manufacture of lubricating oils and greases
9. Synthetic rubber manufacture
10. Coal, oil, wood or nuclear-based thermal power plants
11. Vanaspati, hydrogenated vegetable oils for industrial purposes
12. Sugar mills (white and Khandasari)
13. Craft paper mills
14. Coke oven by-products and coal tar distillation products
15. Alkalies
16. Caustic soda
17. Potash
18. Electrothermal products (artificial abrasives, calcium carbide, etc.)
19. Phosphorous and its compounds
20. Acids and their salts (organic and inorganic)
21. Nitrogen compounds (cyanides, cyanamides and other nitrogen compounds)

22. Explosive (including industrial explosives, detonators and fuses)
23. Phthalic anhydride
24. Processes involving chlorinated hydrocarbon
25. Chlorine, fluorine, bromine, iodine and their compounds
26. Fertilizer industry
27. Paper board and straw boards
28. Synthetics fibres
29. Insecticides, fungicides, herbicides and pesticides (basic manufacture and formulation)
30. Basic drugs
31. Alcohol (industrial or potable)
32. Leather industry including tanning and processing
33. Coke making, coal liquefaction and fuel gas making
34. Fibre glass production and processing
35. Pulpwood, pulp, mechanical or chemical (including dissolving pulp)
36. Pigment dyes and their intermediates
37. Industrial carbons (including graphite electrodes, anodes, midget electrodes, graphite blocks, graphite crucibles, gas carbons, activated carbon, synthetic diamonds, carbon black, channel black and lamp black)
38. Electrochemicals (other than those covered under the alkali group)
39. Paints, enamels and varnishes
40. Polypropylene
41. Poly(vinyl chloride)
42. Cement with vertical shaft kiln technology pending certification of proven technology on pollution control
43. Chlorates, perchlorates and peroxides
44. Polishes
45. Synthetic resin and plastic products

2

Existing Practices: Critiques and Lacunae

The Himalayas, an important geographical region, is cardinal for both plains and hills. It is crucial for the regional climate, biodiversity and the environment. In an industrial perspective, the Himalayan region, which is believed to be a storehouse of natural resources and home to a large variety of minerals, lack research input. Very little research work has been done with reference to mining and mining-related problems, and therefore, authentic literature on the subject is scanty. Thus, this area provides a good scope for research work.

In this chapter, we give a careful analysis of case studies of ongoing mining projects and practices in the region. It is observed that the mining practices described briefly in these case studies have lacunae too. The analysis of the presented case studies and other mines of the region has helped identify the lacunae, thereby paving the way for scientific planning and management.

Various environmental and statutory regulations and social issues are discussed to observe and understand the critiques of mining, enabling the design of an appropriate strategy for Himalayan minerals.

2.1 Case Studies

In Chapter 1, a review of existing mining practices of the Indian Himalayas has revealed that artisanal mining (small-scale mining) has become synonymous with the Himalayas though large mines also exist. According to mineral economists, it is beyond doubt that complete information of mineral availability in the Himalayas is lacking and their mining technique is yet to mature. Two diverging schools of thought have emerged, and hence, before taking any steps for the mining or excavation of minerals, one should consider the fact that the Himalayas are a young cordillera and, according to geologists, the natural mineral formation process through tectonic and mountain-building process is still going on.

Mining being a site-specific activity can be best dealt with through separate case studies. Therefore, operational mines of the Himalayas are described in this section through three different case studies. Limestone, magnesite and soapstone are the minerals covered, and environmental practices at these mines are highlighted.

2.1.1 Kashlog Mine: Mining and Environmental Practices

The Kashlog limestone mine is a large, captive, open-cast, mechanized mine belonging to Ambuja Cements Limited (ACL) and located in the Himalayas (Figure 2.1a). It provides raw materials for the Suli and Rauri cement plants and comprises three blocks, namely the western (Kashlog), eastern (Mangu) and central (Pati) blocks. Mining operations started in May 1994 in its western block, called the Kashlog deposit. In 2015–2016, the Mangu and Pati blocks, which are in close proximity to each other, are being planned and considered for active mining, whereas the Kashlog block is getting exhausted (Figure 2.1b). The mine is situated between the latitudes 31°13′52″ N and 31°15′19″ N and longitudes 76°57′33″ E and 76°59′58″ E and is covered by the Survey of India Toposheet Nos. 53A/15, 53A/16 on the 1:50,000 scale. The mining lease covers an area of 469.00 ha, which falls in the Kashlog, Chakru, Serwala, Banjan, Pati, Banli, Banog, Ratoh, Mangu, Chola, Ghamaru, Serjeri, Gyana, Rauri and Sangohi villages of Tehsil Arki in the Solan district of Himachal Pradesh (India).

Limestone of the lease area is of cement grade, characterized by fine-grained to occasionally medium-grained and hard material. The analysis of the limestone quality indicates that it has a CaO content of 47.27% and SiO_2 (silica) of 10%. The host rocks of limestone in this area are shale dolomite, siltstone and greenish-white quartzite and strata are traversed by joints and fractures. Because of differential erosion, the limestone horizon stands out prominently, while the shale occurs in topographic depressions. Shale formation is also seen between limestone bands in the mine lease area.

Excavation method (method of mining): The mine is being operated by opencast bench slicing from top down. The production statistics is shown in Table 2.1. The deposit has very thin to almost nil overburden cover, and hence, no overburden excavations are being carried out at this mine. The Kashlog deposit can be categorized as having a 'domal structure with outcropping on one flank as massive thickness' (Figure 4.1).

According to ongoing practice, benches of 10 m height are maintained at this mine considering the safety and productivity of the deployed machines. Adequate bench width (30 m) and working space for easy machine movement are maintained. Bench/face slope during working time is 80° to the vertical. Overall, pit slope is maintained below 45°. Currently, there are 24 benches with a working reduced level (RL) of 1565 m (in January 2015). When mining was started in 1994, the topmost RL was ~2100 m at the Kashlog block.

The equipment used include those for drilling (Atlas Copco ROC L8), loading and breaking (PC-1250-7, Komatsu hydraulic excavator-cum-breaker HB-7000), transportation (HD-465-7R, Komatsu dumpers and overland belt conveyor system from mine to plant), compaction (vibratory soil compactor, L&T, 1107D), grading (motor grader, Volvo, G710B) and dozing (D275A dozer and D65E dozer, Komatsu).

(a)

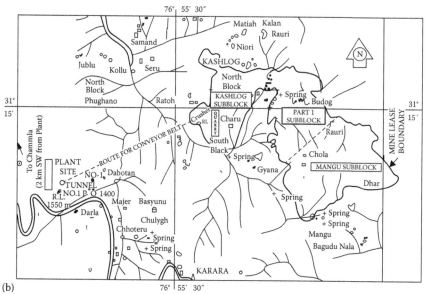

(b)

FIGURE 2.1

(a) Kashlog limestone mine as seen on Google Map. (b) Map showing Kashlog, Mangu and Pati blocks of the Kashlog mine.

TABLE 2.1

Production of Limestone at Kashlog Mine

	Limestone		
Month	2012–2013	2013–2014	2014–2015
April	643,100	565,150	571,200
May	454,900	644,950	596,750
June	486,500	515,500	469,350
July	397,700	336,500	443,450
August	579,100	259,700	442,000
September	407,450	376,950	545,100
October	346,350	375,050	333,200
November	593,850	587,400	540,500
December	452,600	325,150	350,100
January	408,800	437,650	539,750
February	303,650	142,800	377,900
March	413,800	558,450	229,500
Total	5,487,800	5,125,250	5,438,800

Note: All values in tons.

At the Kashlog mine, drilling (150 mm diameter drill) and blasting are carried out through the conventional method in a planned, controlled and scientific manner to minimize fly rock generation for the safety of civil structures, machines and nearby habitation as well as agricultural fields. Due care is taken to keep the ground vibrations and air blast levels to the lowest possible limits to avoid any adverse impacts on the surrounding environment (Figure 2.2). The blasting parameters, such as spacing, burden, depth, subgrade, charge per hole, delay pattern and firing pattern, have been established and predesigned. The non-electric-shock tube initiating system such as EXEL, noiseless trunkline delay (NTD) and Ikon (Digital Electronic System) are regularly used for safety and for reducing noise levels. Loading is done by Komatsu excavators, and hauling of the material is done by dumpers from the mines up to the crusher. From the crusher to the plant, the raw material is transported through closed overland belt conveyors (OLBCs) covering a total length of 2.76 km.

For the Mangu and Pati blocks, another 5.5 km long OLBC combining three belts (OLBCs 1, 2 and 3) is under development and was likely to be commissioned in 2015. No haul road has a gradient steeper than 1:16 at any place except for ramps, where the maximum gradient is 1:10. Adequate drainage has been provided along the roads to prevent erosion due to water run-off. Haul roads have been provided with parapet walls (1 m × 1 m), made of stones or concrete and given 1° slope towards hill drains to avoid the accumulation of water.

FIGURE 2.2
A view of the Kashlog limestone mine.

No blasting is done at the edges or hill slopes. During mining in peripheral areas, big boulders roll down from the hilltop to the valley. To avoid the rolling of limestone boulders towards the slope and also to adopt safe mining practices, the introduction of rock breaker and hydraulic backhoe combination has been done, which is found to be enormously successful. The rock breaker/hydraulic cutting drum is able to break the limestone continuously, and at the same time, the backhoe enables the removal of broken materials by pulling them up towards the benches.

To check the sliding of minor debris and the stabilization of the mine/ road slopes, specific slope stabilization measures and construction of toe/ retaining walls are undertaken at desired locations. The ultimate pit slope is planned to be 45°.

Environmental measures: This is important for mines of the Himalayas and is achieved as per the approved environmental management plan (EMP) of the mine. ACL mine management helps to maintain sustainable development in the entire region including the core zone and buffer zone (10 km radius). The following steps are taken:

Air quality and its management: Drilling, blasting, crushing, laying of haul roads and transportation of the limestones are potential sources of air pollution. A road roller with 10 tons capacity is deployed to compact the loose surface of the haul road. A water sprinkling tanker is deployed to maintain the surface wet so as to avoid dust to become airborne from the haul road. Drilling, another source of dust emissions, is controlled at its generation source adequately by an in-built arrangement of water sprinklers for dust suppression and by a separate dust extraction system. In addition to the aforementioned, blasting, a unit operation of mining, is also a source of dust and air pollution. Water sprinklers have been provided wherever required to mitigate the emissions from the loading and unloading of limestone and shale. The combined action of water sprinklers, the covered conveyer belt, the green belt around the plant, channelization of emissions and installation of a well-designed dust collector, as well as monitoring these setups regularly, enabled mitigation of the air pollution impact considerably.

Water and soil management: Water is not contaminated during limestone mining activity at this mine. Water is pumped/drawn from nearby 'Pazeena Valley/Khud' for domestic, washing and sprinkling purposes. Water used in the workshop is sent to an oil–water separator. The important measures taken to stop the contamination of run-off surface water from the mined-out area and preventing soil erosion are as follows:

- Ensure that proper measures are taken to prevent soil erosion and uncontrolled flow of mine water. Along the haul roads, drains are laid so that run-off may be discharged from the area in a controlled manner.
- Road gradient of 1:16 is maintained as steeper gradient is more prone to soil erosion.
- Check dams/retaining walls are provided to prevent the rollout of boulders and soil erosion (Figure 2.3).
- Plantings are provided in the gullies, below and above the check dams and in the check filter.

FIGURE 2.3
Check dams/retaining walls to prevent scree flow at the Kashlog limestone mine.

- Shrubs are planted all along the slopes towards the down gradient.
- Check filters are built at the culminating points of water courses.

Noise management: Noise monitoring of the area has indicated that the noise levels are within the standards specified under the Indian Environment Protection Act of 1986. Servicing of vehicles and machines is carried out at regular intervals.

The following measures are taken to control the noise pollution and keep the ambient noise levels within limits:

- Secondary blasting is totally avoided by making use of the rock breaker and backhoe.
- Controlled blasting with proper spacing, burden and stemming is done.
- A non-electrical initiation system is utilized.
- Blasting is carried out during favourable atmospheric conditions.
- The prime movers/diesel engines are of proper design and are properly maintained.
- Pollution prevention method is applied; for example, in the case of HEMM, the operator's cabin is insulated and air conditioned.
- Trees are being planted wherever necessary.

Good maintenance and operational practices: The deployed machines are well maintained so that minimum noise pollution occurs, for example provision of closed insulated worker's cabins near the crusher, which is a major source of noise.

Ground vibration and fly rock control: Ground vibrations and fly rocks are caused by blasting operations and are being monitored regularly and periodically using a blast vibration equipment (Minimate DS-567 seismograph). The following mitigation measures are practiced at the Kashlog mine:

- Blasting as per the statuary guidelines is strictly followed.
- Overcharging of holes is avoided.
- The charge/delay is calculated scientifically to keep this menace to a minimum.
- Blasting operations are carried out only during daytime. No night blasting is done.
- Effective stemming of the explosives is done for the blast holes.
- Electronic detonators are used wherever possible.

Solid waste and topsoil management: There is no significant solid waste generation from the mining areas except the topsoil. Topsoil is stripped only from those areas that are disturbed by excavation, filling, road building, etc. The negligible soil over limestone is carefully scrapped and collected manually and stacked separately. The stacking is for a small duration, and the soil thus stacked is used for revegetation/plantation schemes. Before spreading the topsoil, it is ensured that erosion and sedimentation control structures such as diversions, berms and waterways have been established.

Green belt development: The main aim of green belt development (GBD) in the mined area is to restore the premining aesthetic view of the area and the ecological conditions. The GBD scheme broadly covers the following areas:

- Plantation around peripheral portions of the mine and other built-up structures
- Afforestation of barren areas in the mining lease area
- Raising gardens, parks and haul road plantations
- Plantation for the reclamation/rehabilitation of mined-out blocks

ACL has developed its own nursery near the mine area. Afforestation is done in open areas (Figure 2.4), and species are selected in consultation with local forest officials. Planting is done in pits (0.3 m × 0.3 m) by maintaining a spacing of 2.5 m × 2.5 m all around. While selecting the plants, care is taken to select only those that can grow in the climatic conditions of the area and useful for the nearby villagers too. Public participation is also solicited periodically to know the needs of the local people. The pits filled with soil and manure in predetermined proportions are watered. The growing plants

FIGURE 2.4
Afforestation near the Kashlog mine in open areas.

are cared for in the first 3 years, and when sufficiently grown, they are left to nature's will. Plantation or GBD in the mine area is carried out for the following purposes:

- To attenuate noise levels generated from the mine
- To improve the aesthetics of the area
- To trap vehicular emissions and fugitive dust emissions
- To maintain ecological homeostasis
- To prevent soil erosion and protect the natural vegetation

2.1.2 Mining of Soapstone in the Himalayan Hills

The case study of the Chirang soapstone mine of Almora Magnesite Limited (AML), Almora (Uttarakhand), is presented here to give a picture of the existing mining practices in the Indian Himalayas (Figure 2.5) though it is a small-scale mine. Similar to the Chirang soapstone deposit at the Pungar

FIGURE 2.5
A view of Chirang soapstone mine of Almora Magnesite Limited.

valley of Bageshwar district, Uttarakhand, other deposits are also reported in the Himalayas, which occur as small and large lenticular bodies within the carbonate rock sequence, but they are not mined.

Soapstone, also known as steatite or soaprock, is a metamorphic rock. It is largely composed of the mineral talc and rich in magnesium. It is produced in nature by dynamo thermal metamorphism and metasomatism processes, which occur in areas where tectonic plates are subducted. Natural changes by heat and pressure, with influx of fluids, but without melting, lead to the development of soap rocks. Since such conditions exist in the Himalayas, the occurrence of such rocks has been reported.

Soapstone is relatively soft in terms of hardness because of its high talc content. There is no fixed hardness for soapstone because the amount of talc it contains varies widely, from as little as 30%–80%, and therefore, the hardness value on the Mohs hardness scale ranges from 1.0 to 5.5.

Soapstone mineral is found very close to the upper earth surface and is off-white to snow-white in colour. It can be dug out easily using common digging tools and overlain by soil and loose rocks accumulated on the hill slopes. In the Himalayas, it is excavated on a small scale. Paper, cosmetics, detergents, paints and insecticides are the user industries/consumers of soapstone. It has also been used for carving statues for thousands of years.

2.1.2.1 Salient Details of Mine Lease Area

AML started mining in the area during the year 2005–2006. A mining lease of 5.38 ha land near the Chirang Tehsil village and Bageshwar district has been granted by the state government vide government order dated 10.11.2004 in favour of AML for 20 years. The lease area lies between the latitudes 29°52′23″ and 29°52′27″ N and longitudes 79°54′10″ and 79°49′36″ E, and the entire area is divided into 11 blocks. Most of the land covered under the mining lease is private agricultural land. Part of the lease area (0.46 ha) has been surrendered, as several buildings have been constructed by the local villagers, and for safety reasons, mining cannot be carried out over the occupied area.

2.1.2.2 Excavation Method

At Chirang village, soapstone of the white variety (off-white colour) is found in the slope of the hills. The general elevation of the area is 1150–1400 m, and the total deposit lies mainly on the northern and western slopes of the hill. The average thickness of overburden comprising loose soil and rock varies from 0.5 to 3.5 m. The thickness of soapstone mineral body varies from 1 to 6 m.

Hydraulic excavators (backhoe type) of 0.9 m³ bucket capacity powered by 135 hp diesel engines are deployed for overburden handling. The machine removes the overburden 10 m in advance and exposes the soapstone bed. The exposed bed of soapstone is excavated manually to maintain the grade

TABLE 2.2

Soapstone Production at Chirang Mine

Year	Production (MT)	Overburden (MT)
2012–2013	5160	3750
2013–2014	3245	2270
2014–2015	5795	~4500

Note: MT = Metric ton.
Source: Courtesy of Almora Magnesite Limited, Uttarakhand, India.

of the mineral. Wherever the depth of the mining pit does not allow manual excavation of soapstone, the hydraulic excavator is used to dig and cast soapstone on the nearby surface from where labourers further segregate and pack the mineral according to the different grades. As the working face advances by 15 m, after complete excavation of soapstone, backfilling of the overburden on the mined-out pit is concurrently done. The topsoil stored separately is spread again over the mined-out area and the pit is back filled. The annual production of soapstone and overburden removal for past 3 years is given in Table 2.2.

The soapstone is packed in plastic bags (high-density polyethylene bags) and stacked at the stacking yard near the mining pit from where these bags are transported through mules up to the roadhead. Further transportation of the mineral from the roadhead to various end users is done by trucks. As the run of mine (ROM) is segregated and dispatched to various customers directly as per the required grades, there is no ore processing involved in soapstone mining.

All the mineral-bearing areas of soapstone mining are on private agricultural land, and the topsoil is scrapped and stored at safe places before mining starts.

2.1.2.3 Environmental Measures

AML has made sincere efforts to ensure community-friendly soapstone mining at Chirang with minimum environmental impacts. The details are as follows:

Land: Topsoil is stored separately, and after mining and backfilling, it is spread over the mined-out land (backfilled). The mined-out, backfilled and reclaimed land is returned to the respective farmers for taking up agricultural activities as a permanent source of income.

Water: Due care is taken for the protection of natural drainage. In practice, check dams and retaining walls are constructed to control water pollution. Water sprinkling is done on the mule tracks used for soapstone transportation.

FIGURE 2.6
Agriculture field in the mined-out land of the Chirang soapstone mine.

Air: Air pollution is not a problem in this region, as throughout the
    year considerable moisture remains in the overburden strata. The
    soapstone mineral bed, which restricts dust generation during exca-
    vation, also does not pose any air pollution problem.

Plantation of native species like *banjh* (oak) and *cheer* (pine) has been done
at one Panchayat land of the Chirang village. Nearly, 500 plantations have
been raised on the mined-out and reclaimed land through revegetation and
agricultural operations (Figure 2.6).

2.1.2.4 Society

Community-friendly soapstone mining promoted by mine owners and mine
management is practiced in this mining area, keeping societal impacts due
to mining under control. Since the soapstone-bearing land is owned by local
farmers of the Chirang village, the company enters into an agreement with
the respective farmer who allows mining of soapstone in his land under certain terms and conditions. The company involves each farmer in the mining
and other allied activities after necessary training, and thus in the remote
backward hill region of Uttarakhand, the farmer not only obtains a fair com-
pensation for such period but also gets remuneration against his engagement
as a contractor. AML has contributed significantly towards the development

TABLE 2.3

AML's Contribution towards CSR Activities

		Amount Paid by AML (Indian Rupee)			
Year	No. of Local Families Benefitted	Towards Land Compensation	For Ensuring Proper Reclamation of the Mined-Out Land	Towards Remuneration for Participation in Mining and Related Tasks	Total
2012–2013	16	466,385	55,966	4,115,850	4,638,201
2013–2014	24	1,380,980	184,131	7,692,682	9,257,792
2014–2015	22	1,364,321	181,909	4,461,544	6,007,774

Source: Courtesy of Almora Magnesite Limited, Uttarakhand, India.

of basic amenities in and around Chirang village under the corporate social responsibility (CSR) activities. The details of payment made by AML to the local farmers for the last 3 years (Table 2.3) are indicative of this community participatory model of mining.

2.1.3 Magnesite Excavation in the Kumaon Himalayas

Economically extractable deposits of magnesite in Almora, Bageshwar and Pithoragarh districts of Uttarakhand in the central Himalayas are reported in the literature. The Jhiroli deposit lies in the Jhiroli village of Bageshwar district and stretches between the river Kali in the east and the Alaknanda valley in the west and located on the western mid-slopes of the Jhiroli ridge (Figure 1.7). The average strike length of magnesite deposit is 2.9 km. The total established reserves of magnesite in Uttarakhand alone are about 68% of the total national reserves, amounting to about 3.9 million tons.

Magnesite ($MgCO_3$) is used in ferrochrome industries for steelmaking, fertilizers, food processing, mosaic tiles/ceramic manufacture, as well as the electrical/electrode and chemical industry. It has industrial applications in rubber, glass and textile industries as well. MgO/dead burnt magnesite (DBM) is used as the refractory material in steel melting shops as gunning mass. Cattle feed and magnesium sulphate ($MgSO_4$) are also manufactured from DBM.

2.1.3.1 Jhiroli Magnesite Mine

Geologically, the magnesite deposit of Jhiroli of the central Himalayas lies between the North Almora thrust and the Berinag thrust. There are two formations, namely Jaigangad (Rautgara) and Jhiroli (Gangolighat), exposed in the Jhiroli deposit, which are separated by the Billori thrust. The rocks of

FIGURE 2.7
Jhiroli magnesite mine in the Indian Himalaya.

the area form a part of pre-Cambrian calc zone of Pithoragarh. The Jhiroli formation is a carbonate shale sequence and argillo calcareous succession dominated by stromatolite-bearing dolomite and dolomitic limestone with limestone bands and grey shales. The general strike lies NW–SE and the average dip is 45° towards NE. The general elevation of the deposit is 1350–1750 m.

The Jhiroli magnesite mine (latitude between 29°45'30" N and 29°47'30" N; longitude between 79°44' E and 79°46' E) is a 'contour strip mine' belonging to AML (Figure 2.7). The original mining lease was granted to the Uttar Pradesh State Industrial Development Corporation (UPSIDC) in the year 1963 and was further transferred to AML in 1974. The government of Uttarakhand (previously the state of Uttar Pradesh) has granted a second renewal of the lease for an area of 165.086 ha vide the state government order dated 27 June 2014 for a period of 20 years.

AML started mining in the area during the year 1971–1972. The maximum production capacity of the mine is 60,000 tons per annum (Table 2.4). The limiting stripping ratio of the mine is 1:10. Jhiroli, Naini, Billori, Kafligair, Baskhola, Karasburga, Matela, Chanul, Pantgaon, Joshi Gaon, Gaula and Janoti Palari are the surrounding villages comprising the mine lease.

The mine produces different sizes of DBM (sintered magnesite) and calcined magnesite for industrial uses. Raw magnesite ($MgCO_3$) contains silica, alumina, iron oxide, MgO and CaO in different proportions.

TABLE 2.4

Annual Production at Jhiroli Mine

Year	Production (MT)	Overburden (MT)
2012–2013	40,184	296,650
2013–2014	30,152	251,110
2014–2015	33,082	165,130

Source: Courtesy of Almora Magnesite Limited, Uttarakhand, India.

FIGURE 2.8
Jhiroli open-cast mine at the development stage.

2.1.3.1.1 Excavation Method

Mining in the hilly terrain is being done by an open-cast semi-mechanized method. The height and width of the benches are 8 and 18 m, respectively. Overburden is removed by deep hole drilling and blasting, and magnesite is raised by jackhammer drilling and blasting.

The magnesite is dressed and sized at mine pits manually and transported through trucks to the DBM plant located 7.5 km downhill. Calcination and sintering are done in a vertical shaft kiln at the DBM plant where the run of kiln is further sized and packed for dispatches to the respective customers.

Tippers and hydraulic excavators (0.9 m^3 bucket capacity and powered by a 135 hp diesel engine) are deployed for overburden handling. Figure 2.8 shows the developmental stage of the Jhiroli open-cast mine. Overburden comprising poor-grade dolomite is crushed and screened at a mine crushing unit and sold as ballast (gitti) and sand for local building construction work.

2.1.3.1.2 Environmental Protection Measures

In 1989–1990, AML and the G. B. Pant Institute of Himalayan Environment and Development, Almora, undertook a joint project for the restoration of

damaged mined ecosystem based on an EIA/EMP of the mine. The programme was implemented on the field by executing the following tasks
(Thakur, 2008):

1. Preparation of land
2. Afforestation of waste dump
   a. Nursery development
   b. Stabilization of soil by grass
   c. Planting of local free species
   d. Manuring and fertilization
   e. Irrigation
   f. Biological fencing
3. Reviving fauna
4. Control of erosion and sedimentation
5. Control of air pollution in mine and plant areas

The environmental awareness campaign and concerted efforts by the col-
laborating organizations yielded very positive results.

At present, efforts are continuing to minimize environmental impacts on
the ecosystem. Extensive planting of trees has been carried out in the area
(Figure 2.9) for ecorestoration. The topsoil is stored separately and is used for
planting over the dumps. Two dumps have been stabilized, and the planting
of local species has turned these into green spots (Figure 2.10).

Natural drainage is maintained in the mining area as far as feasible, and
check dams, retaining walls, siltation ponds, etc., are being constructed

FIGURE 2.9
Plantation at the Jhiroli mine for ecorestoration.

FIGURE 2.10
Dumps and its restoration at the Jhiroli mine.

FIGURE 2.11
Retaining walls to check scree flow down the slope and to control water pollution.

(Figure 2.11). Common measures of air pollution control, for example regular water sprinkling on the haul roads and crusher area, are undertaken to control dust at the Jhiroli mine.

2.1.3.1.3 CSR Activities by the Mine Management

AML has contributed towards the development of basic amenities in the area under CSR schemes. Local employment to project-affected people (for mining and the setting up of the DBM plant) has been provided, and this has checked and reduced the migration of local population from the backward hilly areas to the developed plains in search of employment. Through local cooperative societies, more than 70 families are being helped by AML to solve their social and economic problems.

2.2 Critique

The United Nations Conference on Environment and Development (UNCED), also known as the 'Earth Summit', which took place in Rio de Janeiro, Brazil, in June 1992 was an important international event 20 years after the Stockholm 1972 Conference, which dealt with mountain areas specifically. Agenda 21 of the Earth Summit formulated an international plan of action for the sustainable development of mountains of the world including the Himalayas. This summit outlined the key policies for achieving sustainable development that meet the needs of the poor hill population and recognized the limits of development to meet global needs. This (Agenda 21) has become the blueprint for sustainability and forms the basis for development strategies in hilly areas. The Rio Principles became instrumental in promoting the development and strengthening of institutional architecture for environmental protection including the conservation and management of mountain resources.

Population analysis world over point out that about 10% of the world's population depends on mountain resources. A much larger percentage draws on other mountain resources, including and especially water. The mountains are a storehouse of biological diversity, future mineral needs and endangered species. For fragile ecosystems with regard to all mountains of the world, two broader programme areas were identified:

1. Generating and strengthening knowledge about the ecology and sustainable development of mountain ecosystems
2. Promoting integrated watershed development and alternative livelihood opportunities

This international agenda gave acceleration to the government, civil society organizations, the private sector and other major groups (hill societies) for recognizing and achieving the goals of sustainable development (UNEP, 2014).

Having reviewed the international agenda of the UN, we tried to find out what was being done in India with respect to the Himalayas. This was essential to bridge the knowledge gap and to formulate a most appropriate strategy for the development and exploitation of mineral resources of the Himalayan region. This scrutiny indicated that subject-wise analysis was available in the literature, but its application to the mining and mineral-bearing areas was missing. For example, the description of the affected environment in hilly areas where mining is carried out is available in the form of technical papers by Indian authors, but how the affected environment can be improved, what significant steps are needed and how to implement them are missing. A number of technical papers consist of introduction and conclusions but no concrete solutions to the issues. We therefore take stock of the situation in the following paragraphs and describe who did what.

Ghose (1993) analyzed the small-scale mining of the region and emphasized on environmental protection but kept the mining agenda above the environment. He emphasized that an EMP, which helps to ensure that the potential environmental impacts a small-scale mine, should be incorporated in the early stages of developmental planning. In another paper, Ghose (2003) has examined in more detail the legislative environment in India and how the performance of resident small-scale mines affected the national mineral output. This paper examined the unique technoeconomic and sociocultural characteristics of selected small-scale mining regions in India including case records of slate mining in the Kangra Valley of Himachal Pradesh. It underscored the need for cleaner production in such regions and outlined a series of legislative measures pertinent to the industry.

A geoenvironmental appraisal of the impacts of limestone mining in Himachal Pradesh by Roy (1993) describes the land use pattern of the Sataun–Dadahu limestone mining areas of the district of Sirmour. Rai (1993) also described the geological and geoenvironmental constraints of mineral resource development in the lesser Himalayan tract of Sikkim and the erstwhile Uttar Pradesh (UP, Himalaya). The areas studied included the hill mining areas exclusively. The author concluded that the overall impacts demand judicious land management but provided no innovative solutions in his research work.

Soni (1994) analyzed the socioeconomic conditions of the inhabitants of the mineral belt of the Himalayas by taking into consideration various factors including land use, land holding and its distribution.

Nagarajan et al. (1994) studied remotely sensed data of limestone mines in and around the Sataun area of Sirmour district, HP. They found from the studies that a significant reduction in vegetation cover (8.89 km²) had occurred, but the agricultural area had increased during the period 1967–1993.

Maynard and Walmsley (1981) described the role of terrain and soil information in the planning of mining development particularly in satisfying many of the existing provincial regulatory requirements. If terrain and soil information is properly used, it can provide a good database for cost-efficient and environmentally acceptable land use for mining areas in hilly topography.

Ding-Quan and Li-Zhong (1989) described some experiences of mining in the high mountain areas above the snow line in China (not the Himalayas). They discussed the features of mining in high mountain deposits, namely development by using the long-pass system, the basic parameters of ore pass, the ventilation of high mountain mines, stability of pit slope and precautions against mountain sickness, as well as the measures taken in Chinese mines.

Mohnot and Dube (1995) dealt with the status of mining in the Himalayas in general and emphasized that there was a need to promote scientific mining in the geodynamically sensitive and fragile mineral-bearing zones of the Himalayas.

Banerjee (1993) addressed the need to combat various problems related to mining in the hilly region. Some general solutions were suggested to ensure eco-balance in hilly areas.

Garg (1990) mapped and monitored the areas affected by mining in the Dehradun–Mussoorie mine belt of the Doon Valley using remote sensing–based methodologies and assessed the impact of mining on the environment. The same study included another hilly area, that is the Kudremukh iron ore mines in Karnataka. Paithankar (1993) also made a case study of the Kudremukh iron ore mine (a mine in hilly terrain but not in the Himalayas) with particular emphasis on the environmental protection measure taken by the mine management. This has helped in understanding the status of hill mining in other parts of the country. Other hill deposits of iron ore, limestone, bauxite, etc., are also being mined by the open-cast method. Such terrain, though identical in topography with those of the Himalayas, does not fall in the category of ecologically fragile zone/areas. Studies by Thakur and Kumar (1993) and Thakur et al. (1992), regarding the management of wastes at altitudes and reclamation of steeply sloping dumps, deal with the Bailadila iron ore mine in hilly areas. In this study, suggestions were given for waste reclamation that are practically applicable in hilly areas. Sen et al. (1991) made a case study of siting the waste dump for an iron ore mine in a hilly topography.

The Darjeeling hill area, the lesser and sub-Himalayan belt of the Sikkim Himalaya, that is the eastern part of the Indian Himalayas lying in between Sikkim towards the north, Bhutan towards the east and Nepal towards the west, is well described by Desai in her book (Desai, 2014), narrating Darjeeling as the queen of the hills. She described the geoenvironmental perception of the entire area as a geomorphologist. The geoecological setup, the natural resource management and its impact on the land use pattern, infrastructure, etc., including human development in the Darjeeling hill area were considered. The people and their culture, economy, agriculture (tea gardens) and tourism were also described, with a chapter on vulnerability assessment and sustainable development of the Darjeeling hill areas. She emphasized that landslides are the major threat to life and property. The economic structure cannot be analyzed merely by observing the towns and roadside development. A house-to-house survey in the rural village areas will reveal the real picture, as it will show how people are struggling for their livelihood. The rural folks/villagers have to walk all day long to sell their produce in the towns. Safe drinking water, educational institutions, primary health services, power supply, transport and communication means face a similar fate in the rural hills. In the recent past, due to the unprecedented growth of population and unplanned development in the hill areas, there has been massive deforestation in the hill areas, which has increased the frequency of landslides and soil erosion. These have become common features nowadays. The pillars of development (education, health, nutrition and employment) are entirely politically driven, and this has resulted in poor development. In her book, Desai (2014) points out mining as one of the causes of land degradation, but beyond that she says nothing about mining. From the geoecological angle, some analysis done by the same author is very good and can be referred to if the Darjeeling hill areas need to be described.

Slopes and their management are one critical issue that affects mining drastically. It has impacts on the scenic beauty of the hills as well. The Alsindi limestone deposit of the Himalayas is a new deposit located in a rugged and difficult-to-access terrain. M/s Lafarge (India) was asked by the Department of Industries, Government of Himachal Pradesh, to explore and exploit this deposit by opening a new mine for limestone production. The company was interested in setting a cement plant in the area, which would also result in the industrial development of the state. As a result of this, the company took up the initiative to scientifically study the slopes in the area.

The lease area of the Alsindi limestone deposit falls under the Karsog Tehsil, Mandi district, in the state of Himachal Pradesh. It is surrounded by the Alsindi, Jankhuni, Talain and Besta villages. The lease area forms a part of the lesser Himalayan ranges characterized by an extremely rugged topography comprising high peaks, steep slopes and deeply incised valleys. No flat or plain land is available in or around this lease area. The mountain that hosts the limestone deposit rises from the banks of the river Sutlej that meanders within the intermountainous valleys. The elevation of river Sutlej is 620 m above the mean sea level (MSL). The highest elevation in the central part of the area is 1910 m above MSL. The relief, therefore, in the buffer zone is 1310 m. The surface plan of the mine area shows that, in general, the hilly topography of the lease area has very steep slopes (>50° and sometimes as high as 75°). The design of pit slopes was therefore essential for this acute angle deposit with rugged slopes. Taking into account the geotechnical parameters of the encountered rock mass, appropriate slope stabilization measures for the Alsindi deposit of the Himalayas were recommended (CSIR-CIMFR, 2014). This scientific study was found to be extremely useful for developmental planning, the preparation of mine plans for mineral extraction, selection of approach routes and location of various amenities and also for important decision making.

Indeed, in many parts of the world, where hilly areas are prominent, artisanal mining (small-scale mining) is ongoing for thousands of years, and it plays a major role in the national context. Minerals and metals for industry, sustainable development of the hill region and development of local opportunities for the local population, particularly in hills, are not new concerns internationally. A number of Asian countries (Nepal, Bhutan, Philippines,

Papua New Guinea, Bangladesh, India, Pakistan, etc.), Canada, New Zealand, Australia, South Africa, etc., are the beneficiaries in terms of revenues generated by mineral production on their lands and the development of backward regions (Soni, 1994a), through local employment and non-migration of indigenous communities to cities, which include mountainous areas as well. Some other teething negative problems for the territories, namely erosion of traditional livelihoods, cultural conflict, loss of land, environmental damage, forced displacement, violence, etc., that can be described as 'side effects', have also cropped up.

In the year 2000, a global/multicountry-level project called the 'Mining, Minerals and Sustainable Development Project' (MMSD), managed by the International Institute for Environment and Development (IIED), London, was started, which convened a dialogue of the major stakeholders on indigenous peoples' interactions with the mining, minerals and metal industries. including the hilly areas of the Himalayas. A country paper from India was included in MMSD (2001) to share the new model of shared, participatory examination of the various problems of Himalayan artisanal mining. The work was helpful for a number of small entrepreneurs engaged in the mining industry and gave directions to the affected parties.

This critique and study of other related/relevant documents has identified the lacunae of hill mining practices, which need further improvement. This is covered in the next section of this chapter.

2.3 Lacunae of Mining and Environmental Practices

With respect to the Himalayan mines and environmental management practices in mining, the following lacunae have been identified.

2.3.1 Land Management Practices and Loss of Vegetation

The process of mineral extraction is bound to cause vegetation loss as the mineral is hidden below the upper layer of the earth. In this process, vegetation in the form of trees, shrubs and grasses is destroyed. Field studies at the mines confirm that in most of the mining areas of the Himalayas, the loss of vegetation differs from area to area. In some places, it is significant, whereas in some others, not much loss is observed. The ancillary activities related to mining also add to the loss of vegetation, for example approach road construction.

The main problem with regard to land management practices and loss of vegetation lies in the adopted practices. There is a lack of proper planning and their actual implementation. The loss of soil and vegetation cover can be reduced to a considerable extent if scientifically devised land management practices are put into practice. The land management methods at various

FIGURE 2.12
Wild cactus in the Himalayas as an ecological indictor plant for limestone.

stages of mining are basically designed for plain areas and are adopted in hilly areas. Hence, the degree of success of these management methods becomes low, and the practices are declared many a times as defunct. This lacuna needs attention for the Himalayan mines.

It should be noted that the limestone mining areas of the Himalayas have excessive growth of wild cactus compared to normal areas (Figure 2.12). The entire hill area (top and slopes) is as such barren. Thus, wild cactus is an 'ecological indicator plant' for limestone occurrences in the Himalayas. This also supports the argument that due to limestone mining, not much vegetation loss occurs and land management is rather easy in such areas. However, scientific land management practices are essential to reduce the vegetation loss to a minimum.

2.3.2 Blasting, Vibrations, Noise and Air Overpressure

Blasting in mines and the vibrations caused by it are environmental hazards for ecologically sensitive areas. Blasting related damages include those to buildings, temporary (kutcha) houses and structures in the vicinity of the blasting sites, in addition to their impacts on wildlife that tends to move away. In small-sized mines of the Himalayas, blasting operations are not scientifically conducted. Scientific ways and means are available that may help reduce the ill effects of blasting and vibrations or may altogether eliminate the blasting process. In extreme situations, vibrations due to blasts can trigger or induce landslides in mountainous regions.

Noise pollution caused by the operation of machines and blasting in hilly areas becomes more pronounced as the sound produced is echoed at the lofty hill structures. Serious efforts are sparingly made to deal with this problem.

Window glass cracking in buildings located near mines is another problem, which is due to air overpressure. All these lacunae of the mine blasting operations, namely vibrations, noise and air overpressure, can be easily dealt with by scientific approaches.

2.3.3 Air Quality and Air Pollution

Air pollution in the context of the hilly Himalayas has two different dimensions. The first is regional and the second is local. In the Himalayas, air quality impacts the entire region through the summer, monsoon and winter season winds, which can transport pollutants (such as black carbon) over long distances; for example, the emission from vehicles in a hill might impact the population in even far away valley(ies) or in remote lower parts of the neighbouring hill country. Thus, local air pollution becomes a regional problem. Cold temperature conditions in the Himalayas lead to more burning of fuel and firewood to keep off the chill, and people knowingly or unknowingly add more pollution to the air and intensify the air pollution problem. This is a local air pollution problem. The mining areas of the Himalayas face both these.

Air quality impacts the environment in such a way that communities experience it every day. As such, air is a dynamic natural entity that is in constant motion, flowing from one place to another through atmospheric circulation systems. Air pollutants cause or intensify fog or haze, which not only leads to health problems but can also reduce crop yield from the lack of sunlight and poor visibility. An increase in winter fog, seen in many parts of the Himalayas, is caused by air pollutants (from sources such as industrial activities, brick kilns and burning firewood). This ultimately leads to an increase in pollutants in the entire hill atmosphere as well as plains down below.

Greenhouse gases and short-lived pollutants can also impact the hill air or atmosphere of the hills. Evidence indicates that, overall, the atmospheric temperature in the Himalayan region has already increased. We can quantify, track and mange it by understanding its negative long-term and short-term impacts.

Air pollution in mining areas is mainly caused by the dust generated during blasting and transportation. In most of the small-sized mines, very little care is taken to reduce dust generation at the source. In the hills, the habitation is usually along the slopes or in the valleys, and the settling of airborne dust in these areas can cause serious problems to the people. Airborne dust also affects the vegetation growth in the valley because leaves of plants, shrubs and trees get coated with dust. Higher wind speed in hills makes the air pollution more widespread.

2.3.4 Disposal of Waste

The solid waste generated as a result of open-cast mining leads to environmental hazards. Because of the shortage of available space, these wastes are dumped in the valley(ies). The flow of large quantities of debris in the valley leads to the blockage of drainage channels, causing problems of water channel blockage and flash floods. Thus, dumping and scree flow along the slope are one common lacuna of hill mining practices.

2.3.5 Hydrological Problems

Another important lacuna of the hill mining practices is a hydrological problem. Drying of perennial sources of water, like springs, wells and unknown water sources of the uphill areas, is a fallout of mining. In hills, though there is no lowering of the water table, rock fissures, fractures and their interconnection cause lowering of the water level or drying of water sources. Each case has therefore to be evaluated separately. As such, the impacts of mining on the hydrological regime cannot be ignored.

Mining operations affect the surface drainage pattern as a result of changes in the topography. Fallen debris obstructs the free flow of water in drainage channels, affecting the availability of water in the concerned drainage basins. In those mines where the drainage is not planned properly, the excavation work or mining operation is hampered.

Water pollution (principally chemical pollution) is a connected hydrological issue that becomes noticeable in case of metalliferous mines or ore excavation. But for other minerals like limestone, soapstone, etc., this is not severe. All the lacunae of hydrological problems, if noticed, should be dealt with only through scientific analysis and implementation procedures for sensitive and fragile hill areas and not through shorter routes.

2.3.6 Open Mine Layout and Its Design

Surface mine layout for bench alignment, orientation, etc., which has been adopted in actual practice, has deficiencies that need to be modified scientifically. Some examples are as follows:

- Planning of faces on either the hanging wall (HW) or foot wall (FW) side of the deposit in a wrong way may cause an increase in the overall length of hill roads.

- Improper orientation of faces during layout planning often results in making the working conditions arduous, for example prominent wind direction and face orientation can make working at the faces comfortable.

- Design of mine faces according to the rock conditions and environmental conditions can reduce future problems, for example rock breakage, sunny faces, mine drainage, etc.

2.3.7 Small-Scale Mining

The scenario of small-scale mining in the Himalayan region is not much different from the overall status of small mines in the rest of the country. Such small-scale mining operations are generally characterized by labour-intensive low-capital situations, arduous working conditions, unplanned exploitation and inadequate environmental protection measures. An example of limestone mining in the Sirmour region and closure of mining ventures in Dehradun–Mussoorie can be quoted, which has raised the eyebrows of local inhabitants and has become a cause of environmental concern. Without giving any justification to the ill effects of mining by privately owned small-scale mining, it can be stated that compared to big mines, although the effects of individually scattered small mines are rather insignificant, the cumulative effects of a cluster of small mines are usually substantial and at times more than the effects of a single big mine.

Adoption of shortcut practices by private entrepreneurs owning small mines is the biggest lacuna of the region.

2.3.8 Problem of Dimension of Mining Lease and Its Period

In small-scale mines of the Himalayan region, most mines have very small leasehold areas of the order of a few hectares only. The mining operations necessitate the setting up of other infrastructures for its proper planning and execution. Therefore, on account of the small lease area size, the very purpose of planning on scientific lines is defeated. Most of the mines have a leasehold area of less than 10 acres. Another problem with the mining lease is its duration. Generally, the lease grant for small mines is for a period of 5 or 10 years only. It hinders long-term planning, which is required to be done at the initial stage itself.

Lease areas granted to some competent entrepreneurs are further sub-divided by the lessee (without prior permission from the competent authority), and patty contracts are awarded for the work of the actual running of the mines, making the total process unsystematic and oriented towards financial returns only. Thus, one single hill slope has more than one mine owner with inadequate lease areas, which is difficult to develop on the scientific lines with systematic mining (Figure 2.13).

2.3.9 Environmental Planning, Practices and Enforcement

Environmental problems due to mining, with particular reference to the Himalayas, are aggravated as a result of the fragility of the system and unscientific approaches. These are the prime causes for the degradation of the major environmental parameters and call for the application of scientific

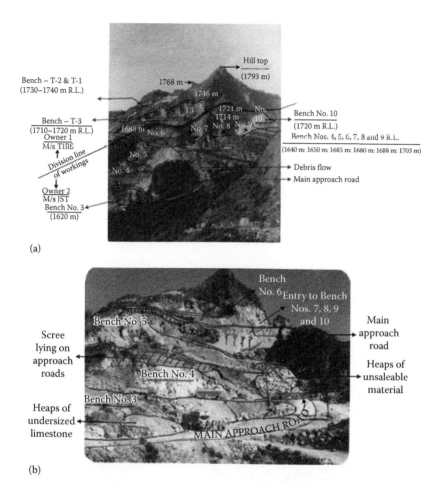

FIGURE 2.13

A Himalayan mining slope with inadequate lease areas and more than one mine owner (a) More than one mine owner on a single slope in the H.P. Himalayas and (b) a slope of the Indian Himalayas depicting inadequate mine lease area. (After Sharma, A.K. et al., Environmental awareness among small scale mines—A case study of Baldhwa Limestone Mine of M/s Jai Singh Thakur and Sons, District Sirmour, Himachal Pradesh, *Proceedings of the National Seminar on Eco-Friendly Mining in Hilly Region and Its Socio-Economic Impacts*, Shimla, India, June 1997, MEAI Himalayan Chapter, pp. 276–285.)

planning and judicious methods for mitigation. Various contributory factors and practices are outlined in the following:

- Lack of efforts for reclamation and ecological rehabilitation of mined-out areas on account of financial constraints
- No attention to profitable end use of mined-out area in the decommissioning phase

Invariably, such important environmental planning and practices, together with social aspects, are not appreciated and are ignored by small mining entrepreneurs. Right from inception to the decommissioning phase of the mine, planning must continue and should end with a final mine closure plan (FMCP). Lacunae of the environmental plan and practices can be best managed by improving the enforcement mechanism. Using an integrated biotechnological approach, if done through a planned approach, for the land management part in particular, will give the best solutions.

All the lacunae and problems narrated earlier are more prominent in small mines of the unorganized private sector. The same problems are handled by large companies more effectively and scientifically. It is also evident from various case studies on Himalayan mining and its lacunae that solutions are available and possible. Extra care and mine-site-specific solutions are needed for the Himalayan hill areas.

2.3.10 Destabilized Hill Slopes/Rock Slopes

In hilly topography, the stability of slopes is a major problem from both the safety and environmental protection points of view (Figure 2.14). Landslides, mudflows, earthflows and other mass waste movements are noticed in hilly areas (Keller, 2011). The main causes identified for the slope failure are improper slope angles and its treatment, removal of the toe of the hillmass, watery conditions, environmental factors and active tectonic forces. In most of the mines, such causes are ignored, and mining practices are implemented without consideration to these factors, creating environmental and operational difficulties for hill slope stability.

FIGURE 2.14
Destabilized rock slopes in a hilly topography (on approach road of a mine; roadside landslide).

TABLE 2.5

Major Landslide Occurrences in the Darjeeling–Sikkim Himalayas (1897–2011)

Year	Area	Causes	Damages
1897	Tindharia and adjoining areas	Earthquake	Infrastructural failures.
1899 (23 and 25 September)	Darjeeling town and adjoining areas	Excessive rainfall (1065.55 m)	Loss of properties and lives (72 persons).
1934 (15 January)	Tindharia, Kalimpong, Darjeeling town, Kurseong	Bihar–Nepal earthquake	Heavy loss of properties.
1950 (14 June)	Kalimpong, Kurseong, Tindharia (Siliguri–Kalimpong railway line was closed forever)	Excessive rainfall (834.10 mm)	Heavy loss of property and lives (127 persons).
1968 (5 and 6 October)	Tindharia, Darjeeling, Kalimpong, Kurseong, NH 31 was severely damaged. Coronation Bridge was washed away (Hill Cart Road especially near Giddah Pahar was badly damaged)	Excessive rainfall (about 640 mm in Kurseong)	Heavy loss of property and lives (67 persons). About 15% of total tea garden areas in Darjeeling was badly affected.
1980 (3 and 4 October)	Darjeeling, Bijanbari, Sukhiapokhri Tindharia, Kurseong	Excessive rainfall (about 300 mm in Kurseong)	Huge loss of properties and lives (250 persons).
1983	Darjeeling town and adjoining areas, Kurseong, Hill Cart Road	Heavy rainfall	Death 5, house damaged 345, heavy loss of property.
1987	Darjeeling, Kurseong, Hill Cart Road	Heavy rainfall	Population affected, 795; death, 17; house damaged, 765.
1997	Darjeeling, Kurseong, Hill Cart Road	Heavy rainfall	Heavy loss of property and lives (death 17).
1998 (7 July)	Darjeeling, Kurseong sub-division and municipality areas (Sherpa busty)	Heavy rainfall	Heavy loss of property and lives (21 persons, more than 3000 houses were damaged).
1999	Darjeeling, Pulbazar, Rangli Rangliot, Simbong Division, Kurseong, NH 55, Kalimpong	Heavy rainfall	Heavy loss of property and lives (11 in Sherpa busty, Kurseong).
2000	Darjeeling–Pulbazar, Kurseong, NH 55	Heavy rainfall	Heavy loss of property but no casualties reported.
2002	Darjeeling–Pulbazar, Kurseong, Tista Bazar areas	Heavy rainfall	Heavy loss of property, total death 7.

(Continued)

TABLE 2.5 (*Continued*)

Major Landslide Occurrences in the Darjeeling–Sikkim Himalayas (1897–2011)

Year	Area	Causes	Damages
2003	Rangli–Rangliot, Tista valley, Kalimpong II, Kurseong, NH 55, Mirik (Gheyabari)	Heavy rainfall	NH-55 near Pagla Jhora was badly damaged, only Gheyabari (Mirik) 23 deaths were reported.
2004	Rangli–Rangliot, Kalimpong II	Heavy rainfall	Heavy loss of property; more than 100,000 people were affected; total death 25.
2006	Rangli–Rangliot, Kurseong NH 55	Heavy rainfall	Heavy loss of property, more than 3000 people were affected and total death 13.
2007	Darjeeling–Pulbazar Kalimpong I, Kurseong, NH 55 Rangli–Rangliot, Sukhiapokhri, Gorubathan	Heavy rainfall	Heavy loss of property especially in tea garden areas like Margaret Hope.
2008	Darjeeling–Pulbazar, Rangli–Rangliot, Kalimpong I, Singringtam, Rangli–Bagogra, Mirik-Karshing Municipal areas	Heavy rainfall	More than 100,000 people were affected but the total death reported was 3.
2009	Sukhiapokhri, Kalimpong I Kalimpong Municipality	Heavy rainfall	About 555 villages were affected; about 150,000 people affected; the total death reported was 40.
2010	Bijanbari, Sukhiapokhri, Darjeeling, Takdah Kurseong, Mirik, Kalimpong I and II NH 55	Heavy rainfall	Total failure of road and rail link in between Siliguri and Darjeeling due to massive landslides near Pagla Jhora and 14th Mile.
2011	Darjeeling, Kurseong, NH 55, Kalimpong	Earthquake and heavy rainfall	Population affected more than 350,000; death reported >90 and houses damaged more than 450,000. Massive landslides occurred on NH 55 near Tindharia.

Source:　Desai, M., *Darjeeling the Queen of Hills: Geo Environmental Perception*, K.P. Bagchi & Company, Kolkata, India, 2014, p. 206.

It is apparent that the destabilization of hill slopes and its causes are many. In the hilly Himalayan areas, whether they are merely natural (rainfall, slope angle/sloppiness, tectonic activities) or anthropogenic (excessive traffic, development of infrastructural facilities, etc.), slope failures are bound to occur. Table 2.5 shows some of the major landslides that had occurred in the hilly areas of the Darjeeling–Sikkim Himalayas, indicating the cause of landslides.

2.4 What Are the Environmental Issues of Concern?

Environment preservation is a significant priority for the Himalayan mountain areas. One single element of mining or the environment alone cannot minimize the negative impacts of development intervention (mining). Mine planners have to analyze the site-specific conditions, both regional and local, to help choose the optimal solutions with respect to mine design and layout. Their priorities are based on engineering judgement on a case-to-case basis, by looking at the site sensitiveness that marks the characteristic feature of the Himalayas. A checklist of environmental concerns is as follows:

- Ambient air quality and air quality at work sites
- Water quality versus hydrological regime of the site and areas nearby
- Noise, vibrations and fly rocks
- Soil and water conservation
- Land degradation including landslides and impact on the aesthetics of the environment (visual impacts)
- Biodiversity preservation and hotspot identification and management
- Flora and fauna
- Ecological fragility
- Waste handling and management including tailings
- Reclamation and rehabilitation
- Wild life (endangered species), archaeology and heritage protection
- Socioeconomic issues
- Subsidence (if, any)
- Earthquake-resistant design for permanent structures involved in hill mining

These identified environmental issues are interconnected or correlated, and hence to address them in totality, a holistic approach is needed.

2.5 Statutory Provisions and Environmental Standards

To date, a number of policies and laws have been formulated to regulate the environmental aspects of India's mining industry as a whole including those in the Himalayas. Many of them are directly relevant to the large mining sector, and some of them are relevant to the small-scale mining sector. Statutory provisions and environmental standards at different mining stages, namely at the exploration (premining), exploitation (mining) and decommissioning (postmining) stage, will help promote cleaner production. Their strict adherence prevents or restricts the environmental pollutants from getting released into the environment, and thus, it is a preventive approach for clean and green mining.

The statutory requirements of the Mines and Minerals (Regulation and Development) Act 1957 (amended in 2015), the Mines Act 1952, the Mineral Conservation and Development Rules (MCDR) 1988, the Mineral Concession Rules (MCR) 1960 (http://mines.nic.in; http://ibm.gov.in), the Environment Protection Act 1986, and the rules, regulations and byelaws framed thereunder are applicable for to the Indian Himalayas (http://moef.nic.in). Similarly, the Industrial Policy of 1991, the National Forest Policy of 1998 and the National Mineral Policy of 2008 are some other statuary frameworks provided under law for the Indian Himalayas. In the mineral sector, various amendments made under these legal tools are also enforceable on hilly territory of the Indian Himalayas for the protection of the environment and preservation of minerals (http://moef.nic.in; http://cpcb.nic.in; GOI, 1991). These policy statements and statutory provisions define the role of the government and lay down plans and procedures for the country's industrial growth. For example, the Policy Statement (1992) for Abatement of Pollution has the following provisions for mining (MOEF, 1992):

1. Mining will not ordinarily be taken up in ecologically fragile areas.
2. A mining plan (as per the MMRD Act 1957) will accompany every mining project, which should include an EMP.
3. It should also include a time-bound reclamation program for controlling environmental damage and restoring mined-out areas according to a progressive mine closure plan (PMCP) and FMCP.

Such general statutory provisions differ from country to country. They are amended, modified and redrafted according to the changed situations, requirements and perspectives of developments. The domain of 'legal provisions and standards' is vast, intricate and equally important as the excavation operation, and its knowledge is essential for smooth operation of hilly area projects.

In a similar manner, industry-specific environmental standards laid down must be adhered to for maintaining adequate environmental quality (http://moef.nic.in). Though plain and hilly areas have sharply different physiographical features, no separate statutory provisions or environmental standards exist in India for the hills. Hence, from a policy perspective, existing laws should be relooked and analyzed to make them more friendly for hilly area applications.

3

Eco-Friendly Perspective
of Himalayan Mining

The Himalayan region, a field laboratory for examining mountain-building processes for the geologists around the world, is distinct in many ways. Its orogeny (geological, geophysical, chemical and tectonic processes of evolution), rugged topography, history and structural features including ores and minerals provide plenty of scope for scientists to learn, for example mineralogical composition of the *collisional ore deposit* found in the Himalayan Mountains, and to get an idea of the chemical and tectonic processes. The *correlative deposits* found in these rocky terrains can provide information, for example, on the minerals likely to be present in other mountainous areas of the world. But here our focus is on exploitation rather than on exploration, and this calls for concentration on the eco-friendly aspects of Himalayan mining in general and with respect to the Indian scenario in particular. To get further details of the eco-friendly perspectives about Himalayan mining, the reader is advised to go through some novel solutions described in Chapter 4 as well. The reading in the consecutive chapters will generate practically applicable ideas in the reader's mind. Having analyzed the Himalayan region and its characteristic features in general, let us switch to the further analysis which prompts us to look at the following pertinent questions:

1. *Is there actual exploitation of minerals in the various other parts of the region including the Indian Himalayas?*

The answer is yes, but only on a limited scale. The entire Hindu Kush region, over which the Himalayas extend, carries out the mining of minerals, but due to the several constraints (extreme and abnormal natural conditions) 'modern mining' is kept to a minimum. To the author's knowledge, this is particularly true in the Nepal Himalayas (http://www.dmgnepal.gov. np) and the Bhutan Himalayas which lack the infrastructure and financial resources, confining mining operations to small/medium scales only.

Bhutan is entirely a Himalayan country and 100% of its minerals lie beneath the Himalayas. The contribution of the mining sector to the gross domestic product of Bhutan in the year 2000 was the lowest among all (1.4% only), that is ~US $1.30 million only. Mineral exploration began only in the early 1960s, and information on the mineral resources of Bhutan is still incomplete. Surveys so far have shown that there are deposits of coal, limestone, dolomite, talc, marble, gypsum, slate, lead, zinc, copper, tungsten, graphite, iron, mica, phosphate, pyrite, asbestos and gold. According to an estimate by the United Nations (1991), reserves of coal (89,000 tons), dolomite (13.4 billion tons) and limestone (121.2 million tons) have been estimated. Most industrial mining of minerals is carried out by privately owned companies operating in the southern part of the Bhutan. Coal is mined on a small scale in Bhangtar in the eastern district of Samdrup Jongkhar and exported to neighbouring tea estates of India and Bangladesh. Currently, the mineral production includes coal, dolomite, gypsum, limestone, marble, quartzite, sand and gravel, slate and talc. For exports, some of the minerals are processed into value-added mineral products such as calcium carbide, cement and ferrosilicon. Most dolomite, gypsum and limestone are mined for the manufacture of calcium carbide and cement. Quartzite is mined for the production of ferrosilicon and microsilica. Most of the mineral production is exported mainly to India and Japan (Roder, 2000).

In the Indian Himalayas, slates in the Kangra valley (Figure 3.1); gypsum in HP and Uttarakhand; sapphire in Sunsan and Paddar (J&K); limestone in Sirmour, Bilaspur and Solan of Himachal Pradesh and Tehri Garhwal area; coal in Assam's hill districts and rock salt in Mandi in Himachal Pradesh are being mined for commercial purposes (Soni, 2003; Soni and Dube, 1995; UNEP, 1997).

The status of mining in the Indian Himalayas further indicates that captive limestone mines of cement plants (in Himachal Pradesh/middle Himalayas) as well as soapstone mines and magnesite mines in the Kumaon Himalayas were operative in 2015. Granite and building stones of commercial grade (in Garhwal and Kumaon Himalayas), which are important building materials, are also mined in the mountainous Himalayan region.

Many companies have tried setting up operations in the Himalayas, but very few have been successful. Some examples of failed ventures are as follows: (1) The Uttarakhand state government had granted license to Pebble Creek Resources Limited, a Canadian mining firm, to explore sulphide ore deposits located in Askot village in Pithoragarh district; (2) the National Mineral Development Corporation, which is a public sector iron ore mining company, had tried limestone mining in Arki of Himachal Pradesh. These companies were unsuccessful on account of many local, regional and commercial issues and could not achieve their desired objectives.

Thus, it is evident that the exploitation of minerals in the entire Himalayan region is going on for many years, but their economic viability in terms of the return on investment is not known correctly.

2. *What is the importance of ecologically fragile hill areas with respect to mineral potential?*

The Himalayan mountains, which fall under the category of fragile hill areas, have plenty of mineral potential. On account of difficult geomining as well

FIGURE 3.1
Manual and age-old method of slate mining in the Kangra valley (HP), Indian Himalayas.
(a) Slate beds and boulders extracted from hills by manual means, (b) waste flow into river
block water channel and transportation by mules, (c) slate dressing and cutting and (d) slate
mining waste generated in dumped hill slopes.

as environmental and socioeconomic conditions, their scientific exploration
and assessment have remained restricted. In case of the Himalayan mountains, both the mountain-building process and the mineral-building process
(the Himalayan region is rising) are continuing, and therefore, uncertainties
dominate the scene. Metallic minerals are found in trace quantities and as
'occurrences', whereas the presence of non-metallic minerals is well reported.
Minerals of the fuel category, although reported, are not yet fully developed.

Since fragile hill areas have high potential for resource degradation even with the normal intensity of resource use, their exploitation beyond threshold limits, imposed by the resistance and resilience of the ecosystem, warrants immediate attention. Adaptation of economic activities (mining) has thus limited scope for resource manipulation to raise productivity levels as well as living standards.

The importance of fragile hill areas can be gauged from the fact that such areas have a series of chain effects on the plains, for example soil erosion and flood in the plains. Disturbances in the Himalayan hills have a triggering effect on the *tarai*. Such impact makes an irreparable dent on the rich and diverse Himalayan ecosystem which needs protection.

3. *What is important for the mountainous Himalayas and Himalayan region?*

This legitimate question is raised to know whether minerals are important for the Himalayas or whether the environment has more significance. Excavation for mining of minerals by digging thousand hectares of land can yield only a few grams of the metal per ton of ores mined. Then, why destroy the precious environment for such a little gain (Jayan, 2004)? The following are extremely important for the mountainous Himalayas and the Himalayan region:

a. The Himalayan region is rich in water resources. The water that flows from the mines or excavated/mined-out areas and flows down the hills into the valley may have pollutants present in it. Water accumulation in the hills can trigger landslides and make the region more prone to landslides.

b. Environmental issues and fragility of the area make the problems more complex. These become more pronounced and important as well because they directly affect the common people.

c. In the Indian Himalayas, lack of specific laws for hill areas separately/alone, lax environmental regulations and their enforcement in the region make the protection of the hill environment a difficult task.

d. Numerous rivers originate from the Himalayas. Any mining activity in the river catchment areas may have a cascading effect on the entire river system, causing pollution, undoubtedly.

e. The Himalayas are a biodiversity hotspot, and the mining industry is notorious for causing environmental degradation. Mining companies, even if they follow eco-friendly practices, are bound to cause environmental damages of the land, water and air. It is not certain whether the positive impacts will be outweighed by negative impacts.

f. Community development is important to fulfill the aspirations of a large section of hill population/society which is by and large poor. Hence, whether commercial exploitation of minerals in the Himalayas by sacrificing environmental concerns is correct or not is a point to ponder. Moreover, in the context of minerals, environment and society, all of which are essential and important for the Himalayas and Himalayan region, a multidisciplinary perspective is required.

Thus, we have answered the question: what is important for the mountainous Himalayas and the Himalayan region? Is it destroying the precious environment or getting minerals from Mother Earth? One has to decide judiciously depending on one's priority.

4. *What should be the approach for the exploitation of mineral resources in the Himalayas?*

It is beyond doubt that any approach for the exploitation of mineral resources in the Himalayas should be environmentally friendly and sustainable. The eco-friendly perspective of the Himalayan mining demands an integrated multidisciplinary approach. The mining should be initiated from the hilltop down and not from the middle of the slope. Mining in the less suitable class (i.e. reserve forest areas, populated areas and agriculture-dominated fertile areas of the valley) may be considered only as an alternative scenario, and in case mining is to be started, proper safeguards should be provided to protect the environment.

Any mining operation should not be looked in isolation but integrated with the environmental component for eco-friendly exploitation of minerals. Therefore, the following points must be paid attention for sustainable development of mineral resources in the Himalayan region.

- Detailed exploration should be carried out for such areas.
- Surface rights should be given with mineral rights.
- Mining should be recognized as an industry (wherever needed) and financing for the necessary mining equipment should be extended to small entrepreneurs.
- Surface right should be given for road construction.
- The mine plan should be systematically implemented.
- Single transport arrangement, in case of two or more lessees on the same single slope, should be made for a group of mines (e.g. skip haulage or glory hole mining method).
- Ultimate pit slope after mining should be less than the natural slope.

- Skilled manpower/competent technical staff should be deputed in the mines, and training facilities for workers should be provided periodically for promoting scientific mining practices.
- Small mines should attempt mechanization collectively.
- Granting of mining lease with particular reference to hill areas should be done keeping in view the fragility and environmental sensitiveness of the area. The lease granting authority should prepare area-specific guidelines for this purpose.
- Mining should not be done in landslide-prone areas.
- Roadside mining (within 50 m) should not be allowed.
- Waste materials from the mines (debris) are sometimes used as resources by local people for house construction, filling and so on. Mine owners sell this waste material, and the government considers it as an economic activity and extracts royalty in some areas. Royalty for debris (if imposed) should be nominal so that they can be used.
- Land management in mines and mining areas should be carried out on the basis of land suitability analysis classes (using GIS and remote sensing techniques).

The mineral resources of the Himalayan region were found to have been exploited after detailed scientific investigations. No short route, that is without proper evaluation and cost–benefit analysis, should be taken. If such practices are not followed, the minerals of the Himalayas should remain as reserve only.

Since environment protection, ecosystems sensitivity (fragility), biodiversity, water, agriculture and food security are on the top of the Himalayan development agenda, an integrated multidisciplinary approach should form the core. The role of mining in the development of backward regions is well recognized (Soni, 1994a), but simultaneously the case histories of the negative impact of mining on regional development should also be taken into account (Soni, 1994b). Taking steps in the direction of mineral sector development, opportunities for the livelihood to the hill people can be provided. It is best if the environment-oriented development of mining areas is carried out (Soni and Dube, 2000).

5. *What is the importance or the role of tectonics in mining and mine design with particular reference to the mountainous Himalayas and Himalayan region?*

The Himalayan region is seismically active and earthquakes occur quite often. In the past, a number of earthquakes (Table 3.1) of different magnitudes have been recorded. Hence, tectonics or seismicity of the Himalayas and Himalayan region (Figure 3.2) is important for mining and mine design.

TABLE 3.1

Earthquakes in the Himalayas

Date	Time (IST)	Location	Latitude	Longitude	Magnitude
25 April 2015	11:41	Lamjung, Nepal	28.147° N	84.708° E	7.8
21 August 2014	13:41	Kangra, Himachal Pradesh	33.1° N	76.4° E	5.0
3 October 2013	11:30	Gangtok, Sikkim	26.1° N	88.7° E	5.2
1 May 2013	12:27	Jammu and Kashmir	33.1° N	75.8° E	5.8
18 September 2011	18:10	Gangtok, Sikkim	27.723° N	88.064° E	6.9
8 October 2005	08:50	Kashmir, India	34.493° N	73.629° E	7.6
29 March 1999	00:35	Chamoli district, Uttarakhand	30.408° N	79.416° E	6.8
20 August 1988	04:40	India–Nepal border	26.755° N	86.616° E	6.3–6.7
20 October 1991	02:53	Uttarkashi, Uttarakhand	30.73° N	78.45° E	7.0
19 January 1975	13:32	Kinnaur, Himachal Pradesh	32.46° N	78.43° E	6.8
15 August 1950	19:22	Arunachal Pradesh (Assam–Tibet earthquake)	28.5° N	96.7° E	8.7
31 May 1935	03:02	Quetta, Pakistan	28.866° N	66.383° E	7.7
15 January 1934	14:13	Nepal–Bihar border	27.55° N	87.09° E	8.2
4 April 1905	01:19	Kangra Valley, Himachal Pradesh	32.01° N	76.03° E	7.8
12 June 1897	15:30	Shillong, Meghalaya	26° N	91° E	8.1

Note: IST, Indian Standard Time; from web sources.

The role of tectonics should be looked at from two different perspectives: (1) in a regional perspective which is related to the continental plates or 'plate tectonics' and (2) in terms of 'neotectonics' which comprise the most recent tectonic movement records available (Box 3.1).

The continent–continent collision between the Indian plate and the Eurasian plate (around 55 million years ago) gave birth to the Himalayas. Despite the long research over the past 150 years, the geometry, kinematics and dynamic evolution of the Himalayan orogeny remain poorly understood (Yin, 2006).

Subsequent to the collision, northward converging 'India' against 'Eurasia' deformed the northern margin of the Indian continent through imbrication of the Indian crust along the major intracrustal thrusts. The topography, geologic structures, earthquakes and resulting landscape response of the Himalayas and the surrounding regions are a consequence of the northward progression and collision of India into Eurasia, a process that has accommodated an estimated 2000–3000 km of convergence since the late Cretaceous age.

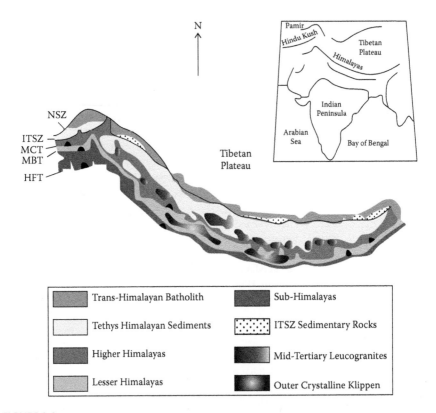

FIGURE 3.2
Map showing the seismic areas of the Himalayas. NSZ, Northern Suture Zone; ITSZ, Indus-Tsangpo Suture Zone; MCT, Main Central Thrust; MBT, Main Boundary Thrust; HFT, Himalayan Frontal Thrust. (From Wilson, l. and B. Wilson, Geology and major structures of the Himalayas, n.d. http://www.geo.arizona.edu/geo5xx/geo527/Himalaya/geology.html.)

While designing a mine or an engineering structure in the tectonically active areas, two types of basic structures are encountered:

1. Natural structures (slopes, mine benches)
2. Man-made structures (buildings, earthquake-resistant designs, permanent structures for life-long use, e.g. machinery installations and reinforcement of various kinds of civil construction)

For these structure types, the suggested solutions are as follows:

1. For the design of natural structures, for example slopes and benches, in a mine in the tectonically active areas:

   a. The overall factor of safety (FOS; for engineering designs) should be taken as 2–2.5 as against 1 or 1.25 in the case of plain areas.

BOX 3.1   TECTONICS

Plate tectonics: The earth's surface consists of a series of tectonic plates, with each plate of the crust and the more rigid part of the upper mantle, termed *lithosphere*. The lithosphere contains all the world's continents and is underlain by a weaker zone called the asthenosphere, with the lithosphere–asthenosphere boundary controlled by temperature. The asthenosphere is solid, but it yields by hot creep, allowing solid-state flow that compensates for the motion of the lithosphere plates. The lithosphere may be more than 100 km thick beneath older continental and oceanic regions, but it is only few kilometres thick beneath mid-ocean ridges where it first forms. Lithosphere plates are in motion with respect to one another, and their movement is called plate tectonics.

Neotectonics: A large part of the continental blocks shows signs of recent crustal instability, and young mountains like the Himalayas is a testimony of it (the Himalayas are rising). This is also manifested in recurrent seismicity, movement of rock masses along faults and thrust planes, subsidence and rise of the ground, structural deformations and evolution of certain peculiar geomorphic features. The imperceptibly slow, secular and episodic crustal movements that have been taking place since the beginning of the Quaternary Period some 2 million years ago are called as neotectonic activities. Areas involved in or subjected to the impact of continental drift and collision are particularly prone to persistent tectonic instability and the attendant deformations and displacements. The Indian subcontinent is one such area of present-day geodynamic movements. The significant and recognizable indicators of neotectonics are historical instances, geomorphic features, structural dislocations and deformations, elevational variations and recurrent seismicity.

Tectonic activities lead to earthquakes, and they trigger natural landslides, avalanches and upliftment of the Himalayas. A mine designer must take into account these factors and keep a sufficient FOS for both man-made and natural structures.

b. The mine planner should consider only those deposits for exploitation that are not near the shear zones, fault zones or any geological discontinuities.

c. The mine designer must take into account country standards, that is earthquake-resistant design of structures (Bureau of Indian Standards; IS 1893–1984; in two parts; Indian Standard Criteria for Earthquake Resistant Design of Structures).

2. For man-made structures (buildings, permanent structures, etc.) in a mine in the tectonically active areas:

   a. Soil–structure interaction: resonance due to matching of frequency

   b. Earthquake-resistant designs

   c. Stiffer foundation design in order to avoid uneven settlement

   d. Provisions of reinforcement in brick masonry especially at weaker sections, that is at corners and joints

   e. Provision of lintel bands (60 cm below the roof, a concrete reinforcement layer)

   f. Expansion joints to take care of expansion and contractions in roof particularly

   g. Symmetrical plans of buildings and structures (square shape rather than rectangular)

   h. Finite element analysis for special structures at the planning stage

   i. Base isolation techniques for foundations of machinery installed in the mines (use of sand in foundation)

6. *In the opinion of an expert (a mining engineer or an alert environmentalist), is it desirable to opt for mining in the Himalaya?*

Mining, as an industrial activity, is a notorious activity against nature. Its negative impacts cannot be kept hidden for long, whatever precautionary steps are taken. Hence, mining in the Himalayas should be limited, selective and controlled.

Whatever the reasons stated, that is technology advancement and research, the first step for opting for mining in the Himalayas should be the economic viability of the mineral deposits, which should be assessed scientifically and before the exploitation begins. As a rational policy, mineral excavation should be restricted for future and only for limited use, and a condition similar to the polymetallic nodules (a mineral form found in large quantities in ocean floors but their present excavation has been restricted on account of environmental reasons) should be followed.

4

Solutions: New, Practical and Eco-Friendly Ideas for the Himalayas

In the previous chapters, we identified the lacunae of mining and environmental practices based on the mining in the Himalayas and its current status in general. Also, based on case studies of working mines in particular, the environmental concerns have been identified. We highlighted the need to intensify the search for the new ideas. We have seen that the conventional mining practices and manual methods of mineral extraction that are in vogue in the Himalayas are less productive and environmentally unfriendly. Therefore, some new restructured ideas to formulate a development strategy for mining complexes in the fragile Himalayas have been framed. These ideas can be easily integrated with the existing conventional methods to make the mining green and conducive to the environment. To cope up with the mining problems likely to arise from geological uncertainties (during mining), advance prediction of geological conditions is desirable in the Himalayas. In this chapter, taking the various ideas one by one, Cases 4.2.1 through 4.2.8 are described and possibly considered as 'solutions' for sustainable mining. These eco-friendly solutions give insights into new and brighter mining concepts that are acceptable and practically feasible for the Himalayas and for the hilly regions in general. Their application areas can be based on engineering judgement, aim of the mining project and the encountered environmental conditions of mine site. The suggested solutions are applicable to various mineral types depending on their economics of extraction, conservation and other related aspects of mining. Tailor-made adjustments need to be made for their implementation and for the selected specific ideas or solutions.

4.1 Predictive Assessment of Geological Conditions

It is well known that the Himalayan terrain has remained underexplored even today. Since Himalayan geology is complex and frequently varying, mining in the Himalayas is risky and uncertain. Prior assessment of the geological conditions at places where mining is planned can reduce this risk and make the mining ventures more attractive and commercially viable. Hence, in this section, we emphasize that prior scientific assessment is essential and desirable both at the exploration and at the mining stages. The reasons why it is required in the Himalayas are as follows:

1. Geological complexities such as thrust, shear zones and folded rock sequences.
2. In-situ stresses and squeezing ground conditions.
3. Vertical and lateral rock covers or overburden.
4. Drilling restrictions in hilly areas and at higher reaches.
5. Ingress of water and gases in the underground openings.
6. Running ground condition while excavating (slope stability and slides for acute angled slopes, etc.).
7. Geothermal gradient (hot water springs) for the higher Himalayas and upper reaches.
8. Engineering challenges in terms of construction methodology, approachability, financial constraints and technical expertise (Mauriya et al., 2010), which are acute in the Himalayas.

When any mining engineering project is commissioned or executed, macro- and microgeological records, feasibility assessment and detailed project report (DPR) based on scientific investigations are needed. On account of inadequate technical information, the investment planned may become a waste. In areas like the Himalayas, technical information based on secondary data is not very helpful, and the regional data may or may not suffice the purpose. In some cases, regional geological data are fine-tuned or extrapolated to get local data to know what exactly lies where. But such details may become misleading many a times. Therefore, primary data are desirable and needed. On account of the severe and rugged topographical constraints, primary site-specific data of the field are cumbersome to obtain and many times impossible also.

To generate primary site data and make prediction of geological conditions in advance, nowadays many advanced methods of investigation are available (Dhawan, 2013). Geological, geophysical and rock mechanics investigations, 3D and 2D numerical modelling, slope stability techniques, advance drilling equipments, etc., are required for this purpose. To collect and collate the primary data, some of the available helpful and advanced tools are as follows:

- *For geological investigations*: Kriging, in situ stress analysis by hydraulic fracturing techniques, computer modelling, etc.
- *For geophysical investigation*: Electrical resistivity tests, seismic reflection and refraction techniques, geophysical tomography techniques, bore hole logging, cross-hole seismic tomography, georadars, etc.
- *For geotechnical and slope stability investigations*: Slope stability radar, 2D and 3D geotechnical and slope modelling using various software and rock mechanical investigations, such as rock mass rating (RMR)/ slope mass rating (SMR), for stability and support assessment.

Scientific investigation of the geological condition in advance can provide sound and well-thought-out solutions for a mining company making investment in the Himalayas. Thus, practical solution for uninterrupted mining lies in the advance prediction of geological conditions and practical applications. Particularly, underground mining, in which lies the future of Himalayan mining, can get its advantage considerably.

4.2 Solutions for Responsible Mining in the Himalayas

In this section some ecofriendly solutions which makes Himalayan mining responsible have been dealt with and analyzed. Eight such case are described below.

CASE 4.2.1 Underground Mining of Limestone

The limestone deposits in the Indian Himalayas (Uttarakhand, Himachal Pradesh and the NE Himalayan region) provide numerous possibilities where underground mining can be adopted. Past precedence (discontinuance of the mining lease in the Doon Valley and acceleration of limestone mining in the adjacent Sirmour region) shows that the principal reason for curbing mining activity is related to environmental protection and preservation. Curbing or stoppage of industrial activity (mining) is not a permanent solution to such problems. Since the best quality limestone is available in the Himalayan region in ample quantity, there exists a need to formulate environmentally benign mining methods in synchronization with the basic principle of sustainable development.

In the Indian Himalayas, domal forms of limestone deposits commonly occur. Three variants are observed: massive bed thickness outcropping on both flanks (Figure 4.1a), limited bed thickness outcropping on both flanks (Figure 4.1b) and limestone beds dipping along the slope/against the slope (Figure 4.1c). Hill deposits with progressive overburden and lying at depth (i.e. deposits dipping into the hills/against the slope) are cheaper to extract by underground means compared to surface methods. Low environmental risk and better conservation of mineral wealth are the key favourable points for the adoption of underground methods in the ecologically sensitive Himalayas. The long-term gains overshadow the short-term drawbacks of the underground method, especially when the question of selection of the appropriate mining method arises. Though production from underground mining is less, the environment and ecology of the Himalayan region can be made much safer by this strategy.

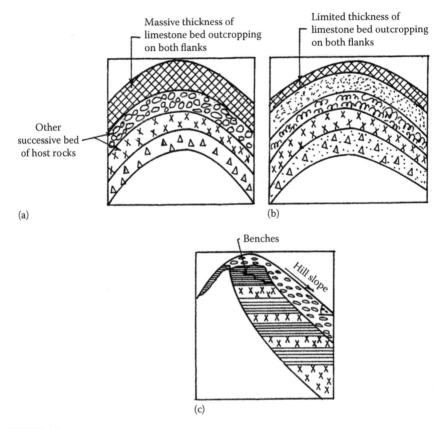

FIGURE 4.1
(a–c) Domal forms of limestone deposits and its variant commonly occurring in the Indian Himalayas.

Deep-seated limestone beds exposed along escarpments cannot be mined by conventional surface mining techniques. For such deposits, underground alternatives are the best. Moreover, the land degradation resulting from surface mining is of a more or less permanent type and visible from a long distance in hilly topography, thereby becoming an eyesore.

Open-cast hill mining leads to the development of wasteland, which is a major cause of concern, nationally. Underground options are the permanent, eco-friendly solutions to land degradation problems, which is a problem in hill mining. Since land is in short supply in the hills, underground methods are definitely superior to surface mining methods. A comparison of the two major methods, namely surface alternate and underground alternate, with specific reference to the limestone mining indicates that the deterring factors are the cost of production and safety. These two factors, especially from the environmental degradation angle, can be debated for a project for the assessment of feasibility. With a scientific approach, the overall mining cost can be reduced and safe practices can be ensured. The negative impact on the environment, which may lead to undesirable and disastrous consequences, may even be repaired permanently.

The author was involved from his institution in a research project called the 'Development of Underground Mining Technique for Limestone Mining in the Western Himalayan Region' and found that the demerits of surface mining can be eliminated by planning and designing a suitable underground method for limestone mining in the Himalayas. This will be certainly a significant step towards the protection of the environment and the sensitive ecology of the Himalayan region.

On reviewing the state of the art of limestone mining worldwide, it is found that most of the underground operations have been developed as an extension of the surface operations when the overburden thickness increases to such an extent that the cost of the stripping operation becomes uneconomical. Underground mining has been adopted in some selected cases to overcome environmental pollution and extreme weather conditions. In several states of the United States, such as Missouri, Iowa, Kentucky, Ohio and Kansas, underground limestone mining is preferred and increasingly adopted as an alternative to surface mining. In Kansas, a number of underground limestone mines have been developed that are still under active production. The biggest underground limestone mine in the United States produces about 1 million tons of limestone (crushed stone) per year, and the deepest underground operation is about 70 m from surface in the state of Ohio (IBM, 1982).

Thus, it is quite clear that underground mining in the Himalayas is a viable and eco-friendly approach at present and in the future if the economics works out favourably for the selected limestone deposit (see Table 4.1).

According to a rough estimate, the cost of underground mining is 25%–75% more compared to open-cast mining. Since underground alternatives are less disturbing to the fragile and sensitive ecosystem and aesthetically, they are more appropriate compared to surface alternatives (open hill/open pit) and, therefore, result in less long-term impact on biodiversity, ecology and environment.

TABLE 4.1

Underground Mining of Limestone

- Limestone beds should be flat and horizontal, and maximum permissible dip should be 1°–2°.
- Geological disturbances, for example faults, shear zones and folds, should be few.
- The bed thickness should be at least 4–5 m.
- The depth of working should be less than 100 m.
- The water problem should be preferably nil or very small.

Underground mining of limestone by the bord and pillar method, access by adit, transport by conventional means or by belt conveyor and ventilation through natural means is feasible.

CASE 4.2.2   Small-Scale Mining: The Concept of Cluster Mining

In general, the large mines draw the attention of the mining and environmental community, whereas small-scale mining (SSM), though common throughout the world, is often ignored. In many countries around the world, SSM is also referred to as 'artisanal mining' and constitutes a major part of the total mining activity. It also contributes towards environmental degradation significantly (Table 4.2). Several most commonly used minerals are mined through SSM, for example sand, gravel, building stones, limestone, gold, coal and gemstones.

Based on literature search, it is found that throughout Asia and Africa several millions of individuals are directly and indirectly involved in SSM (Hilson, 2003). The MMSD project found that undoubtedly the large number of operative SSMs (8000+) in India is causing extensive environmental damage (though not necessarily the worst offender in this regard) (Chakravorty, 2001). In the Himalayas, the environmental concerns of SSM are much higher compared to those of big mines in the organized sector. Therefore, environmental concerns due to SSM cannot be ignored (Hilson, 2003).

Small mines of the Himalayan region depend largely on manual labour, with a significant number of illiterate labourers. Lack of scientific input, guidance and support has led to the development of the SSM sector in a haphazard manner. Therefore, we must find action-oriented solutions so that progress on various aspects of the SSM in different environments can be achieved, as shown in Table 4.3.

TABLE 4.2

Small-Scale Mining and Environmental Impacts

Name of Mineral/Ore	Impact Ranking	Environmental Impact on				
		Water	Land	Human Health	Mine Safety	Scenic Beauty
Building stone	4	Low	Med.	Med.	Med.	Med.
Sand/gravel	5	High	Med.	Med.	Med.	Med.
Limestone	2	Low	High	Med.	Med.	High
Gemstone	6	Med.	Med.	Med.	High	Med.
Gold	3	High	Med.	High	High	Med.
Coal	1	High	High	High	High	High
Others	3	Med.	Med.	Med.	Med.	Med.

Notes: High, medium and low are the relative terms for impact assessment.

Numbers 1–6 given for impact ranking are in descending order of magnitude, where 1 refers to the highest overall impact and the most urgent need for action and 6 refers to the lowest overall impact and the least urgent need for action.

TABLE 4.3

Outlook for Various Aspects of SSM

S. No.	Broad Area of Concern	Action/Solution (Adopt Scientific Approach and Ensure Enforcement)
1.	Environment	*Destruction of vegetation*: Prevent and, if this exists, repair.
		Impact on water: If this exists, reduce and mitigate.
		Impact on air. If this exists, reduce and mitigate.
		Ecology/flora/fauna: Adopt a scientific approach and ensure enforcement.
		Reclamation/rehabilitation: Adopt an integrated biotechnological approach for revegetation (Juwarkar et al., 2008, 2010).
		Seasonal effects: Deal as per the situation scientifically.
		Socioeconomic impact: If this exists, reduce and mitigate.
		Cultural impacts: Deal locally if this exists, but do not ignore.
2.	Mining (refer to Chapters 6 and 7; Sections 6.2, 6.3 and 7.1)	*Production*: Achieve better productivity.
		Mechanization: Increase the use of mechanical equipment and also enhance the degree of mechanization in small units.
		Waste handling: Adopt the best practices in mining.
		Export/import: Harness the potential of export, for example in case of gemstones, slate, granite and dimension stones. Adopt a policy of increased export and reduced import.
		Finances/loan: Make available for mine owners.
		Gender issues: Should not be biased. Non-tribal/tribal issues should be dealt with according to the prevailing cultural scenario.
3.	Health and safety	Perhaps health issues are more aggravated in SSM in recent years despite the advancement in medical facilities, and its proper assessment calls for detailed and extensive investigation all over the region. This should form an important topic of future research and be dealt with in a scientific way.
		SSM operations are mainly manual and hence are more prone to fatal or serious accidents.
4.	Legal	*Government's role*: It should be defined and focused. Practical enforcement should be carried out through periodic inspection.
		Area-specific bye-laws are needed for the hills. Fix rigorous standards.
		Statutes and legal provisions: Comply with and ensure implementation of the existing laws.
5.	Net income	Innovative ways and means must be searched and implemented into practice on a case-by-case basis.
6.	Training	This is needed periodically; hence, implement into practice.
7.	Recognition	*Award/reward*: Recognize and give due importance as per the case cited.

All these recommendations, action and solutions should be critically examined and suitably implemented. This will yield better results for artisanal or SSM in the Himalayas. The common environmental concerns (land degradation, water quality and revegetation) and mining concerns (production planning and scheduling, waste/overburden handling and placement and mechanization) can be effectively dealt with by adopting the best practice mining, as described in Section 6.2.

Small mine owners rarely have the relevant experience like large mining companies. They often remain dependent on the expertise of others (a state mining agency or consultants) to develop plans and implement them. Financial crunch and enforcement are their major constraints. Lack of appropriate technical knowledge and information often put small operators into trouble and add fuel to the fire.

In the Himalayas, a large number of mines are small. Some SSM companies produce very limited quantity of minerals, less than 1000 tons per month. On a single hill slope, where the minerals reside, more than one entrepreneur is engaged in excavation (Figure 2.13) and the leases are allotted to various mine owners for limited time periods. These SSM companies and ancillary operations related to mining become a means of sustained employment for the local population. The local economy is largely controlled by these entrepreneurs, and these mines are mostly operative in the unorganized private sector. Over a period, the mushroom growth of a number of small mines is seen in the various Himalayan mining areas, for example in the Sirmour limestone mining belt of Himachal Pradesh. Such mines operate haphazardly and in an unregulated manner. Enforcing eco-friendly mining operation in these small mines is one of the most challenging tasks for the administration. To regulate them properly on scientific lines, an improved concept of cluster mining has been evolved, which is described in the following section for the SSM companies of the Himalayas. The ill effects of mining operations can be easily controlled and contained by applying this concept.

Cluster mining concept: This is a conventional mining concept treating a number of small mines as one single unit for the purpose of mine planning, environmental planning, infrastructure development and scientific assessment. The number of mines in a cluster has no upper or lower limit but will depend on the size of leased area and local requirements. It should preferably be more than 2. There are two categories of cluster mining: (1) developed naturally during decades of operation and (2) preplanned and executed under the same authority fairly and quickly.

Cluster mining should not be confused with 'isolated small mines' that are widely dispersed in different parts of a country or region. Cluster mining has high employment potential with a high degree of certainty. The concept of small-scale cluster mining as a means of sustained employment and eco-friendly mining operation is nothing but proper pooling of various available resources for betterment. It is a tested concept and able to attract the attention of many discerning personalities connected with mining.

Cluster mining planned with initial technical, administrative and marketing/financial support can yield environmentally friendly results together with safety and productivity at low cost. Because the problems of development of such clusters are so diverse and dominated by many local problems (to be handled technically and socioeconomically), each region/country may need in-depth study and consideration on a case-to-case basis.

When developed with imagination and long-term vision using common engineering sense, it proves not only to be a big source of productive employment but also convenient in shaping the local economy on modern lines. The financial investment needed for generating local employment is also minimal as compared to the creation of high-tech industrial employment. This concept works better and more quickly if the whole operation is preplanned (technically and administratively) by a government body or a resourceful organization with some authority. But the concept and its implementation need to be taken up as a mission with responsibility and not merely as a technobureaucratic exercise which would merely ensure investment but not results. Such efforts would generate confidence in the small entrepreneurs in commercial ventures and develop a breed of small investors. This would be one of the many practical ways of permanently diverting educated unemployed youth from the path of mindless antisocial activities. Cluster mining can be conveniently introduced for *industrial minerals* and *minor minerals* of the Himalayas.

In India, examples of cluster mining do exist in the plains, for example diamond mining cluster in Panna (Madhya Pradesh), marble mining cluster in Rajasthan, glass sand mining in Naini (Uttar Pradesh) and stone aggregate mining in Pachami-Hatgacha in Birbhum (West Bengal) (Chakraboraty, 1997). But, unfortunately, in most of these areas, the total advantage of cluster mining cannot be harvested for want of adequate follow-up and supportive actions by many of the government authorities who are more interested in collecting royalty income than improving the mining activities consistent with safety and protection of the environment. They are not much interested in gradually modernizing mining and processing practices which would improve not only labour wages but also the revenue for the government. Some important advantages of such clusters are as follows:

- Development of common infrastructural facilities at the mine site at minimal unit investment, for example power line and road, railway siding, water supply, telephone and public transport facilities
- Common approach to EIA and EMP on a cost-sharing basis which would be a possible solution to environmental-related problems of isolated small mines, which cumulatively can cause considerable damage without any scope for restoration
- Common geotechnical investigation for the entire area at minimum unit cost

- Common technological, administrative, marketing, accounting and advisory support
- Utilization of costly equipment and machinery on a cost-sharing basis
- Development of common material supply facilities, such as fuel, explosives, cement, steel and common equipment
- Reduction of exploitation by middle man through joint resistance
- Development of a good and viable marketing strategy for mineral sale which is essential for the sustenance of mines
- Reduction in the smuggling of minerals by the checks and balances associated with the cluster mining concept and its implementation

Using certain basic and necessary steps, cluster mining in the Himalayas can be introduced for a planned development. The Sirmour limestone mines and other limestone mines of the Himalayan region, including the closed mines of Doon valley, can benefit directly by this approach. Other private mining hubs/groups, where a significant level of environmental damage is caused by the mining activity, can also take advantage of this strategy.

CASE 4.2.3   Continuous Miner: Experience of Ambuja Cement (at Kashlog) and Lafarge Cement (at Alsindi)

Sufficient raw material production is dependent on proper and efficient mining and the method selected for bulk production. There are four possible excavation methods applicable to surface mining of limestone in the Himalayas: conventional methods of drilling, blasting and loading by shovel dumper combination; method of ripping and dozing; rock breaking by breakers and transportation by trucks or dumpers and production by mechanical means, by a surface miner.

A 'surface miner' is a piece of equipment for continuous surface mining and is meant to extract, crush and load the material in one go. This machine may be an alternative to conventional method of mining. It thus eliminates drilling and blasting and also the primary and secondary crushing in the mining of minerals or rock deposits. Surface miners can even avoid deployment of bulldozers for overburden removal, as bulldozers are equipped with rippers that require regular undercarriage servicing and often generate inconsistent end products. Using surface miners, one can effectively bring the primary crusher up to the face. Thus, a surface miner also serves as an in-pit crusher in another form, and mines can get rid of primary crushing equipment.

The surface miner concept is not new; its origin is from civil engineering applications. Earlier, this mechanical equipment was more or less limited to soil removal, soft land cover digging, soft overburden removal, etc. Later on, its application in mining was extended for surface mining in the removal of softer minerals and materials such as lignite, coal, sand and clay. In the late 1970s and early 1980s, the first improved cutting systems were introduced to the surface mining industry, which were able to cut harder materials also. Different European companies tried to develop machines on the basis of their earlier experiences in the mechanical cutting of rocks. Some noted companies that developed surface miners for mining and other similar application areas are Caterpillar, Dosco and Voest Alpine, Thyssen-Krupp, Wirtgen GmbH, Huron and Morrison Knudsen and Man Takraf and Larsen & Tubro.

Since surface miner technology is environmentally friendly, its applications in the Himalayan mining was thought of, and two past cases were recorded. The first was the deployment of a Wirtgen continuous miner at the Kashlog limestone mine of Ambuja Cements in Himachal Pradesh, and the second was for the feasibility evaluation at the Alsindi deposit of Lafarge.

At the Kashlog limestone mine, the first demo version of the Wirtgen 1900 SM model (middle drum type) was put to use in 1994. This continuous machine became unsuccessful on account of the frequently changing geology and rock hardness. The machine is now deployed at the Kodinar unit of Ambuja Cements in Gujarat. In the case of Alsindi, the hardness, toughness (compressive strength) and abrasiveness of the material were determined in the laboratory prior to deployment of the machine, and it was found that the material of Alsindi deposit cannot be extracted using a continuous miner or some other form of mechanical cutting action. Drilling and blasting are the only feasible solution for the Alsindi deposit (CSIR-CIMFR, 2012).

In the Himalayas, the economics of continuous cutting (or rock fracturing) changes very fast due to varying geology, and therefore, at present machine mining cannot become an alternative to conventional mining. However, predictive assessment has shown that some of the rocks or rock-like materials, which were tough and abrasive once, can be mechanically extracted in the future. Rapid technological advances in hard rock cutting vis-a-vis thrust/force applied will allow increasingly harder and tougher materials to be extracted by continuous miners, but it is not known when. For continuous miners to be competitive, it appears that the rock strength should be less than 80 MPa and the rock material should have low abrasiveness. Thus, an alternative, environmentally friendly option for the Himalayan mining is machine excavation using a continuous miner, but the material to be excavated must be tested beforehand.

CASE 4.2.4   Innovative Transportation System
in Use in the Indian Himalayas

An innovative system for ore transportation from the hilltop to the bottom is implemented and adopted in practice at the Gagal limestone mine of M/s ACC Limited (now ACC is owned and managed by M/s Holcim Limited). This is called a vertical silo with conveyor arrangement (Figure 4.2). For an open-cast mine of 1 MT or more run of mine production per annum, the daily handling of mineral/rock is of the order of 3500 tons per day. In such cases, transportation by cylindrical bins and conveyor belt is the easy and cost-effective alternative. The cylindrical bins of steel, also termed as *surface silos*, act as the hollow shafts and can be erected vertically on the slopes by providing steel support frames.

Figure 4.2 shows a schematic representation of the cylindrical bin and conveyor arrangement. The layout is suitable for moderate production requirement and preferred by the companies also, as they are easy to install. Vertical silos can be made of thick steel sheets and can be fabricated in a local engineering workshop. To make the transportation more effective, the crusher location can be decided at the hilltop and the advantage of gravity transportation can be taken. The whole transportation distance is covered in stages (steps) by conveyors C1, C2, C3, C4, etc., depending on the conveyor length and capacity. Arrangements for chutes, feeders, etc., can be done as per the design of the installation.

The advantages of this cylindrical silo conveyor transport arrangement are its simplicity, ease of installation and low maintenance cost. Continuous and uninterrupted production can be achieved with this kind of arrangement, and the system is applicable for any hill locality.

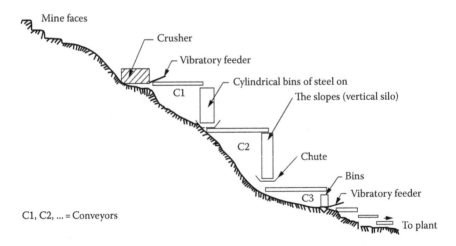

FIGURE 4.2
Eco-friendly ore transportation in hills (cylindrical bin and conveyor arrangement).

CASE 4.2.5 Slope Hoisting System and High Angle Conveying

When surface mining goes deeper and the mine takes the shape of a funnel, transportation of the ore from the pit bottom to the pit top becomes costlier. Also, fuel consumption and vehicular maintenance plus wear and tear increase considerably. The deeper the mine funnel, the greater will be the expense for transport. At the same time, air pollution due to dumper or truck transport also increases. To optimize transport while maintaining the flexibility offered by truck transport, the Mining Technology Division of Siemag Transplant Germany (GmbH) has developed a 'trucklift slope hoisting system' which considerably accelerates and reduces transportation cost from the mine (Figure 4.3). This system has applicability in hilly terrains for mineral/ore transportation from the valley level to the hilltop level when mineral deposit lies in a deep valley. Rugged terrain conditions can be easily coped with, and a good degree of environmental protection can be achieved.

Other obvious advantages of its implementation are the curtailment of transport time, savings in fuel and reduction in air pollution. The system can be designed for a depth of 300 m, and multistage lifting is also possible, overcoming the height difference for the lifting trucks. Though its initial cost is high, from the environment point of view, this proposition seems to be very attractive. Its choice and selection will depend on individual cases and the economics involved. Safety has to be ascertained with the highest degree of accuracy for using this system. Hilltop, hill bottom and mechanical conveyance arrangement for truck lifting must be designed with 100% safety factor.

High angle conveyance for hills: Snake and snake sandwich belt conveyor: In the hills, gravity advantage can be taken for the transportation of materials from the hilltop to the bottom. While conveying ore or mineral in

FIGURE 4.3
Trucklift slope hoisting system for surface mines located in valleys. (Courtesy of SIEMAG, Transplan, GmbH.)

bulk uphill and at a steep gradient, the biggest problem is that of slippage on the gradient. Using 'snake conveyors', conveying up the gradient, that is 60° or steeper gradient, is now possible (Figure 4.4). Two systems, namely the 'snake conveyor' and the 'snake sandwich belt conveyor system', can lift large quantities of bulk materials, including coarse products such as rocks, aggregates or coal, at the steepest possible inclines. Dos Santos of the United States have conceived, researched and developed the engineering design for these systems.

What is a sandwich belt conveyor? A sandwich belt conveyor uses two conveyor belts, face to face, to gently but firmly hold the product being carried, thereby making steep incline and even vertical-lift runs easily achievable. Snake conveyors are available in a wide range of profiles (e.g. C and S shapes) (http://www.dossantosintl.com/).

Snake sandwich belt conveyors are reliable as well as economical solutions to steep angle and vertical high angle conveying requirements. They offer the following *advantages*:

1. Unlimited conveying capacity
2. Continuous flow of material
3. Applicable for most rugged mine applications
4. Suitable for friable materials
5. Low operating and maintenance costs by using conventional conveyor hardware, so usable with conveyor system

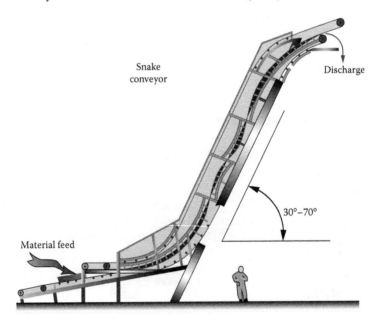

FIGURE 4.4
Snake conveyor for surface mines (for transportation from valley to hilltop).

They are used for conveying bulk materials through 80° or more and are successfully operating throughout the world.

These described engineering transportation solutions are a part of the integrated strategy and best practice mining approach for planned development of fragile Himalayan mining areas and provide a holistic solution. Taking into account the cost analysis, as well as the environmental advantages offered by the new and innovative ideas given earlier, the fragile Himalayan environment can be protected.

CASE 4.2.6 Rock Salt Excavation by Solution Mining

Mining of rock salt has been done in the Himachal Pradesh Mandi salt deposit for the past 200 years and more. This deposit is located in the Indian Himalayas and was in good demand sometime back for local needs, religious purposes and medicinal purposes and as cattle lick. Currently, due to lack of production and quality, this is mostly used as cattle lick only.

Rock salt contains about 65%–75% sodium chloride (Table 4.4), and the rest is insoluble impurities mostly. The minimum content of NaCl in salt should be 96%, and since the rock salt produced from Mandi mines contains less than this, it is not suitable for edible purposes in its raw form; however, processing, if done, can make it fit for human consumption.

TABLE 4.4

Technical Specifications of Rock Salt at Mandi, Himachal Pradesh

S. No.	Name of the Chemical Constituent	Content (% by Weight)
1.	Sodium chloride (as NaCl)	67.81
2.	Insoluble matter	31.23
3.	Calcium bicarbonate	0.35
4.	Magnesium sulphate	0.04
5.	Calcium sulphate	0.26
6.	Sodium sulphate	0.31

As per the United Nations Framework Classification system, the total deposit of rock salt in the Mandi district of Himachal Pradesh is estimated at 16.03 MT (as on 4 January 2010). Out of this, about 10.04 MT is proven reserves, while 5.99 MT is probable reserves (IBM, 2014).

Mining: M/s Hindustan Salt Limited (HSL), a government of India enterprise, and its subsidiary Sambhar Salts Limited have their own rock salt mines at Mandi in Himachal Pradesh. Two prominent mines of rock salt, namely the Drang Mine and the Gumma Mine, are reported operational in HP Himalayas. Drang is located on the Mandi–Palampur road and can also be approached from the Mandi–Pathankot national highway. The villages Bijani and Narla are close to the Drang mine. The Gumma mine is an underground mine for rock salt excavation, where the bord and pillar method of mining is practiced (IBM, 2009). The arduous working conditions and unscientific approach mark the mining operation of rock salt.

Rock salt is produced in chunks and mined by the manual dry mining method through 'pick and axe' mining. On average, about 400–500 tons per month of rock salt is produced from these two mines of the Mandi district. It is mostly used as a natural cattle lick by livestock owners within 300–400 km radius of the mines. Super washed and refined variety of rock salt is packed in 1 kg polypacks and sold in the local market. Because the maximum rock salt (in its raw form) mined is used for local consumption, dry mining by manual means only is economical and therefore practiced at present in these mines. The excavation has not proven very profitable for the obvious reason that rock salt is a low-value mineral. The uneconomical extraction of impurities from the raw rock salt deposits is yet another reason for its non-lucrative extraction. Therefore, to make its exploitation attractive, there is a need to adopt a better, more eco-friendly and viable method of extraction and also chemical processing for quality enhancement.

In this context, it is significant to note that Mandi rock salt has suitable constituents to make it usable for medicinal purposes. Higher domestic and industrial demands for rock salt in the entire northern India can be easily met by these mines if a proper production method is adopted. The salt produced from 'saturated brine' and mined by 'solution mining' is a concentrate and has vast scope even for the export. For solution mining methods to become viable (feasible), demand from urban consumption centres is a prerequisite. This demand should be perennial and not necessarily from the local rock salt producing areas only. By matching the transportation cost from far-off salt producing areas to the consuming areas, its economic viability can be established.

Current status: Rock salt mining in the Himalayas has passed through various ups and downs in the last several years since its extraction. A brief summary of events is given as follows:

> In 2011, the government of Himachal Pradesh barred M/s HSL from carrying out mining activities because of complaints of illegal mining received by the government. In 2011–2012, no production of salt took

place in the mines of Mandi district due to the temporary closure of mining operations. Thereafter, HSL discussed the issue with the state Industries Department for a fresh mining lease and permission to operate the Drang and Gumma mines. In the middle of 2014, Himachal Pradesh government stepped up efforts to restart the rock salt mines by setting up a large plant in Mandi district. The public sector company HSL is yet to apply for the lease grant renewal for which formalities like clearances under Forest Conservation Act, environment clearance and mining plan have to be completed by the company. HSL proposed a plant at an investment of around Rs 30,000,000 for the extraction of rock salt from the Drang and Gumma mines of Mandi district. The HP government issued a letter of intent in March 2014 to HSL. The state government is seeking the central government's intervention for revival, funding and speeding up the rock salt project. The mine authorities approached the HP state government in 2015 for fresh mining leases from the Department of Industries. Only sporadic mining activities are going on at present.

Future mining of rock salt: As described earlier, scientific and economical extraction of rock salt is possible by solution mining which involves dissolving/leaching the rock salt in-situ to obtain concentrated brine and re-extracting salt by evaporating the brine by solar/thermal evaporation. This is achieved by drilling bore holes/wells into the salt deposit, injecting fresh water into the wells, leaching the salt and recycling the weak brine to obtain enriched brine from the salt cavern. The resultant brine is then carried to a salt refinery for obtaining salt that can be used for human consumption or for industrial purposes.

Solution mining is an eco-friendly process and has advantages over dry mining especially in hilly regions and more particularly to the Mandi salt deposit due to reasons of safety, economy and the complex geology of the terrain. After considering all the aspects, it is suggested to exploit the Mandi salt deposit by the brine well method of solution mining, a brief description of which is given in the following paragraphs.

Brine well method of solution mining: In this method, a cased vertical hole of 15–20 cm diameter is drilled from the ground level to a relatively great depth up to the deposit's bottom. For a brine well, a good cover of cap rock is essential. The interface between the cap rock and the salt should be generally well defined. In the Drang salt deposit, there is no hard and compact cap rock, so a good thickness of about 50 m of rock salt has to be left above the cavern roof for supporting the upper strata.

The first and foremost condition for planning the solution mining method is that the geology and other technical data of the brine well or the exploratory well should be well known. Old or previously known data of the salt deposit or well hole data (secondary data) can be useful but not sufficient enough for the purpose of solution mining. To implement the rock salt leaching concept, the rock solubility characteristics and physical well data are essential. Since these data frequently change when the face changes in the leaching area and during the course of the solution mining phase, a planned scientific approach is necessary. This concept should be utilized with all caverns to be developed for extraction.

Solution mining has two methods for the development of cavern (Malhotra, 1997):

1. *Direct solution mining*: In this method, fresh water is injected through the inner tube, and the brine is drained off via the annular space between the inner and outer casings (Figure 4.5b).

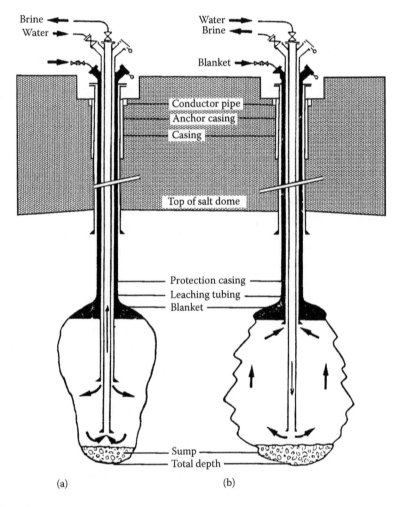

FIGURE 4.5
(a, b) Solution mining concept for rock salt extraction.

2. *Indirect solution mining*: In this case, fresh water is injected via the annular space between the inner and outer casings, and the brine drained off via the inner casing (Figure 4.5a). The proposed solution mining operations are stipulated to be carried out to a depth of 100–400 m from the surface.

The configuration of each cavern (well) must be calculated by a 3D simulation model. The model predicts the exact development of the solution mining process based on the operational variables such as leaching rate, depth of tubing and position of the protective oil blanket. The leaching of the cavern is carried out in different phases.

Leaching the sump: In an impure salt deposit, it is not possible to start solution mining by directly enlarging well to be wet mined. In the Mandi deposit, because the insolubles are more than 30%, a larger cavity sump must be leached first by direct solution mining to allow these insolubles to settle to the bottom of the sump. The sump should be leached to the final diameter of the well (Malhotra, 1997).

Leaching the chimney: A thin chimney is sometimes developed during sump leaching for eliminating the deflection of well. Such leaching is followed by chimney leaching by the direct method. In this phase, a part of the bore hole within the salt mass is enlarged. In the case of indirect leaching without a chimney, the well can get plugged below the point of fresh water injection cavity roof.

Cavity roof: The shaping of roof is done by indirect solution mining. Rock mechanics specialists recommends a cone-shaped roof. However, the roof cannot always be shaped as desired due to geological circumstances. The conical cavity roof is shaped by pumping exact quantities of the blanket medium into the cavern at certain intervals.

Final leaching: When the roof is fully shaped, the cavern reaches its boundary diameter in its topmost region. The water tubing is then lowered, and the upper region is filled with the blanket. Subsequent stages of reverse leaching are continued. The cavern grows large according to the specified volume curve which can be determined from the leaching behaviour of the cavern.

In principle, the number of solution mining steps varies greatly. Therefore, it is important to know the depth at which the tubings are to be set, the moment when the tubing depths are to be changed and pumping the right quantity of blanket medium at the right time.

Based on the discussions, it is concluded that mining of rock salt is possible by solution mining in the future. For the planning and development of brine production fields in solution mining, detailed considerations of rock mechanics, drilling, geology and metallurgical engineering (leaching) are imperative. These can be known from the exploratory well(s).

CASE 4.2.7   Hill Areas Study Using Satellite and Collateral Data

The synoptic view provided by remote sensing satellites offers a techno-logically accurate and appropriate method for studying the ecological and environmental parameters. Well-known and established methods and tools, namely remote sensing techniques using GIS software/Bhuvan portal of the government of India, are the most effective instruments for defor-estation studies, hotspot determination, biological assessment, soil erosion measurement, land degradation analysis, water pollution, agricultural changes, etc. These tools can be easily applied to study large-scale mining, SSM operations and mine environment studies in the Himalayas.

Nearly, all hill areas are tough to approach, and some areas are totally unapproachable and rugged; the condition is similar as that of the Himalayan terrain. For such areas, remote sensing techniques are best suited.

To deal with the various region-specific problems faced by the hill people of the Himalayas, remote sensing-based integrated studies not only are helpful but also provide most viable solutions for the manage-ment and conservation of precious natural mineral resources. Especially for the developmental planning of hill areas, such studies also become necessary because of factors such as altitude, rugged slope and terrain, low literacy rates, wasteland, man-to-land ratio, power and infrastruc-tural facilities. In the case of a group of mines (or cluster of small-scale mines), an integrated approach is immensely beneficial and becomes a

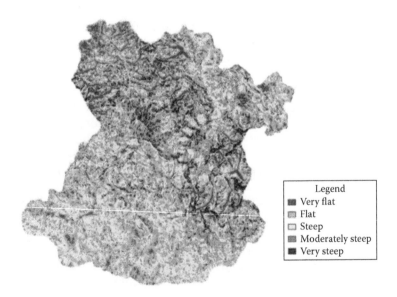

Legend
■ Very flat
▨ Flat
□ Steep
▧ Moderately steep
■ Very steep

FIGURE 4.6
Slope map of the Himalayas using GIS. (After Chakraborty, E. et al., *Int. J. Res. Eng. Technol.*, 3, 141, 2014.)

good basis for inventorying, management and easy monitoring of the hill areas (Gupta, 1992).

False colour composites and multidate Landsat thematic mapper data on 1:50,000 scale can be used in the environmental and planning studies for the preparation of various thematic maps, namely forest map, land use/land cover map, wasteland map, lineament map, slope map (Figure 4.6), watershed map, drainage map, route optimization map (Figure 4.7), ground-water potential zone, demographical profile and density map (Nagarajan et al., 1994). The information so generated can then be used for preparing scientific microlevel development plan or hazard zonation mapping of that particular area and its mineral resources (Chakraborty et al., 2014).

The mineral-rich areas have the potential for industrial development, and in this context, collateral/supporting data, such as rainfall, topo maps, socioeconomic data and developmental problems of selected region, can be easily and accurately identified; also, forest resource infor-mation, current status of land resource use (through change detection analysis), etc., can be updated conveniently (Inamdar et al., 1992).

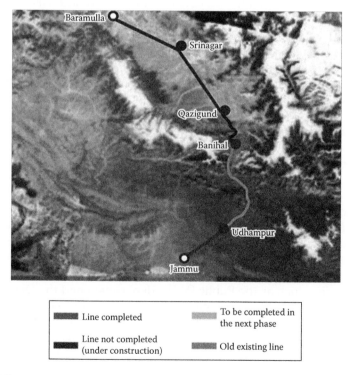

FIGURE 4.7
Satellite image of an optimized rail route in the Indian Himalayas. (After Tejal, P. et al., *Int. J. Curr. Res. Acad. Rev.*, 2, 167, 2014.)

From Case 4.2.7, it is evident that in the Himalayan mining areas modern remote sensing and GIS tools can become useful for developing mines in an eco-friendly manner. As such, these approaches, which form a part of developmental planning, will cost less and yield more. They can be easily clubbed with an integrated strategy which is stressed upon and emphasized here for implementation.

CASE 4.2.8 Slope Failure Management Using Local Resources

Slope failures are common in hills and mines in the Himalayan areas. Both 'roadside failure' and 'mine faces failure' are caused by a number of well-known reasons. Retaining walls are commonly constructed as protection against such failures because they pose hindrance for the normal working condition. Depending on the type of failure and its place of occurrence, management measures are selected and applied. In the Himalayas, slope failures are the major worries for the management handling it, as it involves considerable cost. The management therefore searches for techniques that are cost-effective. Hence, for slope management, innovative solutions are essential that can reduce cost, improve the speed of construction and promote the utilization of slope waste material/local material to the extent practicable.

In managing failed slopes, the transportation of material from far away places and consumption of other materials in construction make the management measures costly. Considering these points, the Council of Scientific and Industrial Research (CSIR), India, and its constituent laboratories that work on engineering problems of slope failure and management have come out with slope failure management solutions using local resources that will be easily available at the slope failure sites. The 'drum debris retaining wall' developed by Central Building Research Institute (CBRI), Roorkee, India (a constituent of the CSIR), is a step forward in this direction. This is an alternate solution to the *Gabion wall* and other reinforced earth constructions.

The *drum debris retaining wall* makes use of slope waste, slide debris, colluvium, talus and other excavated material available at the slope failure site in the construction of retaining walls to stabilize the slopes. Empty bitumen drums available in abundance through road construction agencies, such as the Public Works department and the Directorate General Border Roads, can be interconnected vertically and laterally and anchored at the floor using long bolts (studs) to form a wall-like structure (Figure 4.8). This acts as a 'diaphragm wall' to obtain the desired level of slope stability. The developing agency (CBRI) had applied this technique for the management of one big natural slide, named the Kaliasur landslide, on the Rishikesh–Badrinath road in the Garhwal Himalayas in 1986, and its results indicated that this system was very cost-effective

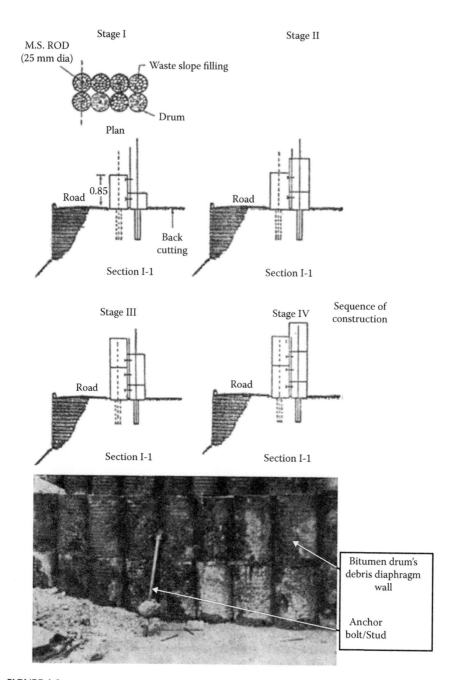

FIGURE 4.8
The drum debris retaining wall for slope stabilization.

and stood the onslaught of medium to high slope movement adequately (Bhandari, 1987).

The *drum debris retaining wall* is an innovative and well-tested system to bring down the expenditure on slope control measure to a minimum and can be used in mines on any road in the Himalayas or near the mine faces. Another advantage is that the long-distance transportation of construction material is eliminated, and tons of wasteful debris usually lost in the valley could be effectively utilized.

5

Environmental Attributes: Two Niche Areas

The Himalayas, a visibly distinct and ecologically important geographical region, are unique in terms of resources, elevation/relief, population, socio-economic characteristics (especially with respect to poverty indicators), bio-diversity and hydropower potential (water provision). At the same time, the Himalayan region is sensitive, fragile and also tectonically active. To address various environmental problems of mining operations in a practical and easy way, adequate methods are required. In this chapter, we analyze two niche areas: assessment of environmental deterioration in quantitative terms and environmentally friendly transportation practices in hilly mines. This is done in detail, as these areas have ample scope for environmental protection. It is observed that though the mine management collects a huge amount of environmental data, it remains practically unaware, how to utilize it effectively for environmental management. The explanation given in this chapter will provide the reader with an opportunity for strengthening his or her ideas and strategies towards practical implementation.

If the environmental deterioration by developmental activities can be quantified, especially due to mineral exploitation, the industrial activity can be better planned and made more environmentally benign. Since all ecosystems develop a set of specialized and integrated structural and functional characteristics, which are set in delicate balance, one can preserve and protect the ecosystems effectively if their intricacies are well understood. The same is the case with the Himalayan environment. Based on research work, this chapter makes an attempt to quantify the mining environment of the studied Himalayan mine area. This is described and explained in detail in Part I.

5.1 Part I: Environment Degradation Index

The assessment of the deterioration of environmental quality can be done by an empirical approach easily and conveniently. Since this method is easy and practically applicable for the quantification of environmental deterioration, it can be adopted for the Himalayan areas. A degradation scale called the *environmental degradation index* (EDI), ranging from 0 to 1, has therefore been formulated utilizing air and water quality field data and remotely sensed land data for assessing the quality of the mine environment. The applicability of this environmental scale was verified in four limestone mines: Manal, Kashlog, Lambidhar and Gagal mines located in the Himalayas. Values closer to 0 are considered environmentally safe. A correlation is also established between the degradation index and the environmental protection cost (EPC), which is being applied or used as regulatory measures for preserving the Himalayan environment and ecology. The details are described in the following sections.

5.1.1 Environmental Indices

The measurement of environmental deterioration in quantitative terms is a difficult task because of the complexity involved. An environmental index is an easy and possible solution for such quantification (Abassi et al., 2012; Ott, 1978; www.csir.co.za). Moreover, environmental deterioration and its quantification (assessment) can be made by developing critical impact factors, also called 'environmental indices' (Canter, 1996).

Three physical attributes of the environment, namely air, water and land, are among the most important components for the preservation of ecology and the ecosystem, as they are generally taken into account while evaluating the environmental status in an area. A number of researchers and authors have developed indices for different application areas, but only a few attempts have been made to develop an index system for the evaluation of mining sites by combining the various mine environmental pollution–related parameters. This is the first ever attempt in the development of an index, taking into account air, water and land pollution parameters (variables), which is referred to here as the EDI. Since it is based on the field environmental data of the Himalayan mines, it will be directly applicable for such areas.

5.1.2 Development of the EDI

An EDI is a representative indicator for the environmental quality of the particular site for which it was prepared. To develop an index, a large volume of environmental field data is statistically analyzed and converted into a compact scale or index. In the process of reduction of representative field data, the derived compact information should not seriously distort the scenario that the data represent. The developed index should be compact and accommodative for future requirements as well.

As such, there are no fixed methods to evaluate the environmental condition of any particular site in which an industrial operation is going on. Therefore, different empirical or analytical procedures according to the requirements are followed (Figure 5.1).

Attempts have been made to quantify environmental deterioration, such as noise annoyance and land use zoning (Sol et al., 1995), coal mine water pollution (Sundarjan et al., 1994), river water pollution (Bhargava, 1983) and water quality assessment of village well water (Krishna et al., 1995). Different

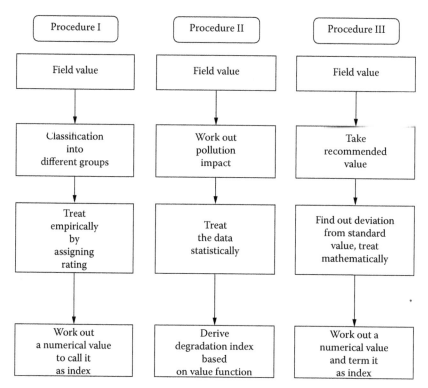

FIGURE 5.1
Various procedures for EDI determination.

empirical or analytical procedures are followed by different authors (Burn and Yultiani, 2001; Daylami et al., 2010; Kleynhans, 1996) for designing indices. To develop and design an EDI, procedure III, as outlined in Figure 5.1, has been adopted. The selection of procedure III has been made mainly because of its simplicity and field applicability compared to other methods.

The EDI developed by the author for fragile Himalayan mining areas is a true representative of the environmental quality because of the fact that it makes use of real field data and as such there is no distortion of the environmental scenario. As the data change, the index also changes.

Since the developed index gives the combined effect of all the three parameters (i.e. air, water and land) in a single numerical value, it will be easier to use it in the decision-making process. It should be added here that earlier attempts to design a degradation index did not include all the three physical attributes (i.e. air, water and land) which are at least required (minimum essential requirement) for evaluating the environmental status of an area.

The pollutants present in air and water are measured in terms of various pollution parameters. These pollution parameters are referred to as 'variables'. According to the Central Pollution Control Board, India (CPCB, 1990), different

TABLE 5.1

Air Quality Data for the Manal Mine

S. No.	Parameter	Air Sampling Data Results									
		A1	A2	A3	A4	A5	A6	A7	A8	A9	A10
1.	SPM	417.6	427.6	318.66	470.66	654.15	393.65	229.72	489.08	714.94	354.17
2.	SO$_2$	1.2	0.73	0	0	<6.0	<6.0	<6.0	<6.0	<6.0	<6.0
3.	NO$_x$	4.7	16.06	18.1	12.3	<6.0	<6.0	<6.0	<6.0	<6.0	<6.0
4.	CO	N.D.	N.D.	N.D.	N.D.	NIL	NIL	NIL	NIL	89.0	NIL

Source: Cement Corporation of India (CCI), New Delhi, India.

Notes: All values are expressed in μg/m^3 unless otherwise stated. N.D. refers to 'not detected' and NIL means detected but not traceable. For A1, A2, A3, …, refer Figure 5.6. In this table, SO$_2$ values of <6.0 are not detectable by the measuring instrument. Therefore, for the purpose of computation, it is assumed that <6.0 means 5.9.

limits of suspended particulate matter (SPM) or PM$_{10}$/PM$_{2.5}$, sulphur dioxide (SO$_2$), carbon monoxide (CO) and oxides of nitrogen (NO$_x$) have been specified or prescribed for environmentally sensitive areas (fragile areas), industrial areas and residential areas in respect of the air quality (Tables 5.A.1 and 5.A.2). Similarly, for water quality assessment, there are 35 constituent parameters, for example pH, total suspended solids (TSs), dissolved solids (DSs), BOD and COD (Table 5.A.3). Each parameter has an impact significance and contributes to the overall quality of the environment.

Considering the importance of the different parameters and their significance with reference to a particular use (Soni, 1997) at a particular mine site, data on air, water and land have been collected and used in the analytical work (Tables 5.1, 5.2 and 5.5).

To develop an EDI, which represents the environmental degradation of a particular site, the first step is the determination of the sensitivity function, index variable and index number, followed by a detailed procedure of calculation as given in the following sections.

5.1.2.1 Sensitivity Function

The sensitivity functions $f(x)$, which are based on the maximum and minimum permissible limits and recommended values of various environmental degradation parameters, are as given in Equations 5.1 through 5.3. The three terms mentioned as the upper limit (Ul)#, the lower limit (Ll)# and the recommended value (rv)# in these equations are as per Indian standards prescribed by the Central Pollution Control Board (CPCB, 1990). These limits are the statutory requirements under the provisions of the Environment Protection Act and Rules. The Bureau of Indian Standards, which is the national standardization organization in India, also prescribes the same maximum and minimum values for various environmental variables for a particular application (Goel and Sharma, 1996).

TABLE 5.2

Water Quality Data for the Manal Mine

S. No.	Parameter	Unit	Pre-Monsoon Period						Post-Monsoon Period					
			W1	W2	W3	W4	W5	W6	W1	W2	W3	W4	W5	W6
1.	pH value	–	7.62	7.79	7.90	7.98	7.85	7.95	8.1	8.2	8.0	8.2	8.1	8.2
2.	Hardness	mg/L	205	195	300	400	375	190	135	170	300	300	250	140
3.	Chloride	mg/L	10	3.0	4.0	3.0	4.0	5.0	10.0	10.0	9.0	10.0	20.0	9.0
4.	Total solids	mg/L	276	181	359	642	448	355	496	274	340	654	684	459.0
5.	Dissolved solids (DS)	mg/L	248	131	340	475	260	265	429.42	234	268	482.4	604.5	346
6.	Suspended solids (TSS)	mg/L	28	50	19	167	188	90	66.58	40.0	51.0	211.6	80.5	113.0
7.	Sulphate	mg/L	15.6	N.D.	N.D.	3.0	198	52.6	N.D.	64.36	54.65	62.40	69.89	54.335
8.	Fe content	ppm	1.3	0.5	0.5	0.7	0.20	0.5	N.D.	0.102	0.008	0.009	0.013	0.064
9.	Ca content	ppm	54.6	42.5	865	97.41	96.2	46.2	14.0	4.2	3.8	3.5	5.9	28.0
10.	K content	ppm	19.2	10.4	10.6	13.1	23.9	56.3	9.23	9.66	27.4	39.4	31.7	15.3
11.	Mg content	ppm	7.83	7.94	13.00	32.60	32.90	11.50	38.86	52.15	75.06	55.38	48.01	30.91
12.	Na content	ppm	31.6	49.2	53.4	58.7	75.8	198.4	23.0	10.0	11.0	23.0	16.5	68.0

Source: Cement Corporation of India (CCI), New Delhi, India.

Notes: N.D. refers to 'not detected' and NIL means detected but not traceable. For W1, W2, W3, …, refer to Figure 5.6.

$$f(x) = \frac{P_{fv} - P_{rv}}{P_{rv} - P_{LI}} \quad \text{where } P_{LI} \leq P_{fv} < P_{rv} \tag{5.1}$$

$$f(x) = 0 \quad \text{where } P_{fv} = P_{rv} \tag{5.2}$$

$$f(x) = \frac{P_{fv} - P_{rv}}{P_{rv} - P_{UI}} \quad \text{where } P_{rv} \leq P_{fv} \leq P_{UI} \tag{5.3}$$

where
  P_{fv} is the field value of the parameter
  P_{rv} is the recommended value of the parameter
  P_{UI} is the upper limit of the parameter as prescribed in the standard
  P_{LI} is the lower limit of the parameter as prescribed in the standard

(*Note:* Here, the prescribed statutory limits fixed by Indian law [Indian National Standards] have been used in the equations, as the studied mines are located in the Indian Himalayas. No confusion should exist as regards the use of these parametric values. These values may differ from country to country and can be considered accordingly for EDI calculation.)

5.1.2.2 Index Variable and Index Number

The index variable (X_1, X_2, X_3, X_4, ...) for each pollution parameter with respect to air and water can be defined as

$$X_n = 1 - \mod f(x) \tag{5.4}$$

and the index number (IN) is given by

$$\text{IN} = \sum f\{X_1 \times X_2 \times X_3 ... X_n\}^{1/n} \tag{5.5}$$

where n is the number of pollutant parameters taken into consideration.

5.1.2.3 Procedure for Calculation of EDI

The following procedure is adopted for EDI determination.

- Three broad environmental attributes, namely air, water and land, are selected.
- Air and water quality data of each mine (or evaluation site) are collected (Tables 5.1 and 5.2) and are taken as the field values for the determination of $f(x)$. Figures 5.5 and 5.6 show the location of mine site and air quality/water quality sampling points, respectively.

- Sensitivity functions $f(x)$ using Equations 5.1 through 5.3 for both air quality and water quality are determined (Tables 5.3 and 5.4).
- The mod value of $f(x)$ is taken and indicated as P in Tables 5.3 and 5.4.
- The index variable X is calculated using Equation 5.4.
- The geometrical mean of index variables using Equation 5.5 gives a numerical value, which is the required *air quality* or *water quality index number* (AQIN or WQIN) for each location (Tables 5.3 and 5.4).
- The average of AQIN or WQIN for more than two location's data of a mine site (or a mining area) gives a composite value for each mine. These multiple locations (core zone and buffer zone) are chosen in such a way that they represent the environmental status of the whole mining area.
- To determine the *land quality index number* (LQIN), leasehold and degraded areas are required. These areas can be determined

TABLE 5.3

$f(x)$ Values for Air Quality Data of the Manal Mine

S. No.	Parameter	$f(x)$ Values									
		1	2	3	4	5	6	7	8	9	10
1.	SPM $f(x)$[a]	3.176	3.27	2.186	3.70	5.54	2.93	1.29	3.89	6.14	2.541
	P[a]	3.176	−3.27	2.186	3.70	5.54	2.93	1.29	3.89	6.14	2.541
	X_1[a]	−2.176	−2.27	−1.186	−2.70	−4.54	−1.93	−0.29	−2.89	−5.14	−1.541
2.	SO$_2$ $f(x)$	−0.959	−0.975	−	−	−0.803	−0.803	0.803	−0.803	−0.803	−0.803
	P	0.959	0.975	−	−	0.803	0.803	0.803	0.803	0.803	0.803
	X_2	0.41	0.025	−	−	0.197	0.197	0.197	0.197	0.197	0.197
3.	NO$_x$ $f(x)$	−0.843	−0.464	−0.396	−0.59	−0.803	−0.803	0.803	−0.803	−0.803	−0.803
	P	0.843	0.464	0.396	0.59	0.803	0.803	0.803	0.803	0.803	0.803
	X_3	0.157	0.536	0.603	0.41	0.197	0.197	0.197	0.197	0.197	0.197
4.	CO $f(x)$	−	−	−	−	−	−	−	−	−0.911	−
	P	−	−	−	−	−	−	−	0.911	−	−
	X_4	−	−	−	−	−	−	−	−	0.089	−
X		−0.1400	−0.030	−0.715	−1.107	−0.1762	−0.075	−0.0112	−0.112	−0.0177	−0.0598
AQIN =		−0.519	−0.312	−0.846	−1.052	−0.561	−0.4215	−0.224	−0.482	−0.365	−0.391

[a] The values are computed.

Notes: $X = (X_1 \times X_2 \times X_3 \dots X_n)$ and $AQIN = (X_1 \times X_2 \times X_3 \dots X_n)_{1/n}$. n is the number of parameters taken into consideration.

AQIN for the Manal mine (average) = 0.519 + 0.312 + 0.846 + 1.052 + 0.561 + 0.4215 + 0.224 + 0.482 + 0.365 + 0.391.

AQIN = −0.51735 (total).

TABLE 5.4

f(x) Values for Water Quality Data of the Manal Mine

S. No.	Parameter		$f(x)$ Values					
		1	2	3	4	5	6
1.	pH $f(x)$[a]	0.573/−0.430	0.6633/−0.4975	0.6333/−0.475	0.7266/−0.545	0.65/−0.4875	0.7166/−0.5375
	P[a]	0.573/0.430	0.6633/−0.4975	0.6333/0.475	0.7266/0.545	0.65/0.4875	0.7166/0.5375
	X_1[a]	0.247/0.57	0.3367/0.5025	0.3667/0.525	0.2734/0.455	0.35/0.5125	0.2834/0.4625
2.	DS $f(x)$	−0.8387	−0.9130	−0.8552	−0.7720	−0.7941	−0.8545
	P	0.8387	0.9130	0.8552	0.7720	0.7941	0.8545
	X_2	0.1612	0.086	0.1447	0.2279	0.2058	0.1455
3.	TSS $f(x)$	−0.5271	−0.55	−0.65	0.893	0.3425	0.015
	P	0.5271	0.55	0.65	0.893	0.3425	0.015
	X_3	0.4729	0.45	0.35	0.107	0.6575	0.985
4.	BOD	N.D.	N.D.	N.D.	N.D.	N.D.	N.D.
5.	COD	N.D.	N.D.	N.D.	N.D.	N.D.	N.D.
6.	SO$_4$ $f(x)$	−0.9844	−0.9356	−0.9453	−0.9673	−0.8660	−0.9465
	P	0.9844	0.9356	0.9453	0.9673	0.8660	0.9465
	X_4	0.0156	0.0644	0.0547	0.0327	0.1339	0.05347
7.	Cl $f(x)$	−0.99	−0.9935	−0.9935	−0.9935	−0.988	−0.993
	P	0.99	0.9935	0.9935	0.9935	0.988	0.993
	X_5	0.01	0.0065	0.0065	0.0065	0.012	0.007
8.	As	N.D.	N.D.	N.D.	N.D.	N.D.	N.D.
9.	Fe	−	−	−	−	−	−
10.	Hg	N.D.	N.D.	N.D.	N.D.	N.D.	N.D.
11.	Ni	N.D.	N.D.	N.D.	N.D.	N.D.	N.D.

(Continued)

TABLE 5.4 (Continued)

f(x) Values for Water Quality Data of the Manal Mine

S. No.	Parameter	f(x) Values					
		1	2	3	4	5	6
13.	Zn	N.D.	N.D.	N.D.	N.D.	N.D.	N.D.
14.	Cu	N.D.	N.D.	N.D.	N.D.	N.D.	N.D.
15.	Fluoride	N.D.	N.D.	N.D.	N.D.	N.D.	N.D.
16.	Phosphate	N.D.	N.D.	N.D.	N.D.	N.D.	N.D.
17.	Cd	N.D.	N.D.	N.D.	N.D.	N.D.	N.D.
18.	Ca	–	–	–	–	–	–
19.	K	–	–	–	–	–	–
20.	Mg	–	–	–	–	–	–
21.	Na	–	–	–	–	–	–
X		2.937×10^{-6}	545×10^{-6}	6.6031×10^{-6}	1.4171×10^{-6}	7.609×10^{-5}	1.5202×10^{-5}
		6.7785×10^{-6}	8.1404×10^{-6}	9.4536×10^{-6}	2.3583×10^{-6}	1.143×10^{-4}	2.481×10^{-5}
WQIN		0.0783	0.0886	0.09203	0.06765	0.01501	0.1087
		0.0925	0.09596	0.0988	0.0749	0.1627	0.1199

[a] These values are computed.

Notes: $X = (X_1 \times X_2 \times X_3 \dots X_n)$ and $WQIN = (X_1 \times X_2 \times X_3 \dots X_n)_{1/n}$.

n is the number of parameters taken into consideration.

WQIN for the Manal mine (average) = $(0.0783 + 0.0886 + 0.09203 + 0.06765 + 0.1501 + 0.1087)/6 = 0.079 (0.0925 + 0.09596 + 0.0988 + 0.0749 + 0.1627 + 0.1199)/6 = 0.10746.$

Therefore, average = 0.09323.

WQIN = 0.079/0.1074 (total).

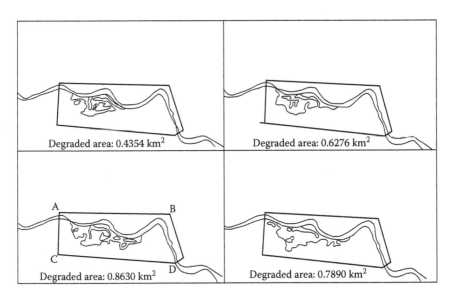

Degraded area: 0.4354 km^2

Degraded area: 0.6276 km^2

Degraded area: 0.8630 km^2

Degraded area: 0.7890 km^2

FIGURE 5.2
Land degradation map for the Manal mine (based on satellite data interpretation).

either by manual mapping of the area using survey methods or from satellite data. Here, satellite data (IRS 1-C) procured from the National Remote Sensing Agency, Hyderabad, are used for the degraded area mapping (Figure 5.2). Degraded area (B) for more than two different time periods (preferably 3 or 4) must be calculated because B versus A graph is to be plotted, considering premining/initial mining land data as baseline. The area in this study is determined using a planimeter (Soni, 1997). Thereafter, the ratio of the degraded area (B) to the lease hold area (A) is calculated (Table 5.5).

- The ratio of B/A is plotted against time. The statistical treatment thus given yields a best-fit line (line XY in Figure 5.3). Equation of this best-fit line can be represented linearly, quadratically or logarithmically. Among this, the linear best-fit line ($Y = mX + C$) is selected. In this regression equation, the value of m can be taken as LQIN. The more the data, the better will be the LQIN.

- The average value obtained by the summation of AQIN, WQIN and LQIN is taken as the EDI.

5.1.2.4 EDI Scale

An EDI scale from 0 to 1 (Figure 5.4) is used to quantify the degradation of air, water and land quality (combined) due to the mining operation.

TABLE 5.5

LQIN Data for the Manal Mine

Lease Area (A) (km²)	Degraded Area (B) (km²)	%	B/A Ratio	Duration (Years)
6.1778	0.1922	3.1	0.031	5.8
	0.4246	6.9	0.069	8.1
	0.3536	5.7	0.057	9.8
	0.2354	–	–	2.5
	0.1614	–	–	4.0

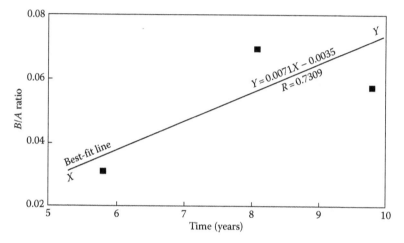

FIGURE 5.3

B/A versus time plot for the determination of LQIN.

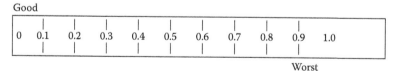

(0) means the environment is good and (1) means the environment is worst

FIGURE 5.4

EDI scale.

5.1.3 Site of Study

To evaluate the environment, one operative mine, namely the Manal Mine of M/s Cement Corporation of India (CCI), a public sector undertaking, was selected, which is located in the Himalayas (Figure 5.5). The Manal limestone mine produces 600 tons of limestone per day (tpd) to meet the requirement of CCI's Rajban Cement Plant (Figure 5.5). The Manal limestone deposit, which is under active extraction, is located on the right bank of the river Giri, the major water body in the immediate vicinity of mine and situated at about 11 km from its cement plant at Rajban. The other details of study site are as follows:

Type of mine	Captive open-cast mine of a cement plant (Rajban Plant)
Latitude and longitude of the mine	30°34′ N and 77°37′ E
Location details (district, state and country)	Sirmaur, Himachal Pradesh, India
Nearest town, railway station and state highway	Paonta Sahib, Dehradun, Paonta–Shillai state highway No. SH-22
Topography	Hilly
Deposit under extraction	Deposit No. 3
Method of mining	Surface method of mining in benches
Rated capacity of the mine	2.8 MT/year
Status of mechanization	Semimechanized
Expected life of mine	83 years (at the rated capacity)
Operational since	1976
Mine lease area	617.78 ha

For collecting the air quality data, different monitoring locations within and outside the lease area were selected. These points/locations referred to as A1, A2, A3, A4, etc. (Figure 5.6) were selected in such a way as to cover both the core and buffer zone radius (5 and 10 mm, respectively) around the mining operational area. For each sampling station, data for three seasons, that is summer (May–June), monsoon (July–August) and winter (November–December), were collected and averaged.

Similarly, for water quality assessment, water samples of one season each before and after the monsoon ('before monsoon' means month of June and 'after monsoon' refers to December) were collected as shown in Table 5.2. In Figure 5.7, the locations of various sampling points are shown and marked as W1, W2, W3, W4, W5 and W6. The sampling point locations for water were finalized based on the study of the drainage pattern and hydrological features of the study area.

Considering the importance of the different parameters (variables) in limestone mining, the values of Fe, Ca, K, Mg and Na obtained from the analysis (Table 5.2) were omitted for $f(x)$ value determination in Table 5.4 (Soni, 1997). In this study, the primary data as well as secondary data (collected from the source organization) were analyzed for estimating EDI.

The field data obtained (fv) over three different seasons and from various sampling points (A1, A2, … and W1, W2, …), for both air quality and water quality differ considerably. Such variation is quite obvious because the local conditions influence the parameter values; for example, the $PM_{2.5}$ or PM_{10}

Kashlog Mine (Darlaghat, Bilaspur)
Gagal Mine (Barmana, Bilaspur)
Lambidhar Mine (Mussoorie, Dehradun)
Manal Mine (Paonta Sahib, Nahan, Sirmour)
Rajban Cement Plant, CCI

SHIMLA

⊙ DELHI

FIGURE 5.5
Location of mines selected for investigation. *Note*: Names of places in parentheses show the nearest towns and district headquarters, respectively.

values within the core zone (or near the mining operation) will differ from the values in the buffer zone. The selection of the sampling points was done carefully and elaborately in such a way that the data (primary data or secondary data) covered the entire mine area, giving a true representation of the mining operation and its influence on the environment.

The computation of EDI for the Manal mine was done by following the methodology described in Section 5.3.1.

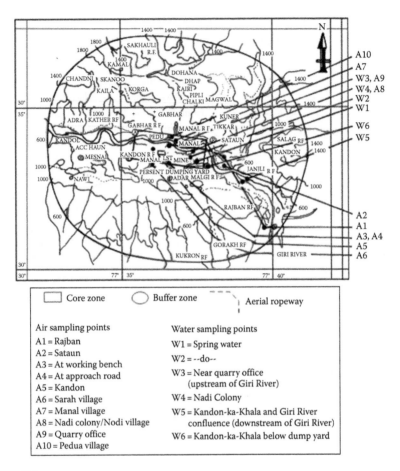

FIGURE 5.6
Location of air and water sampling points for the Manal mine.

EDI computation for the Manal mine

From Table 5.3: $AQIN = (-)0.51735$
From Table 5.4: $WQIN = 0.079/0.1074$
From Figure 5.3: $LQIN = 0.0071$ $(Y = 0.0071X - 0.0035)$

Therefore, *EDI* -0.51735 -0.51735
                      $+0.079$ $+0.1074$
                      $+0.0071$ $+0.0071$
                      ─────────   ─────────
                      0.43125 $(-)0.40285$

Average *EDI* = $(-)0.41705$

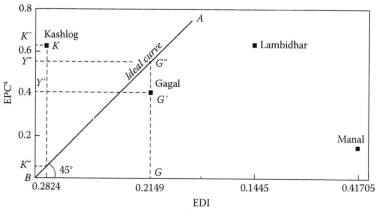

FIGURE 5.7
EDI versus environment protection cost.

TABLE 5.6

Calculated EDI Values for the Studied Mines

S. No.	Mine Name	Calculated EDI
1.	Gagal Mine	0.2149
2.	Kashlog Mine	0.2824
3.	Lambidhar Mine	0.1445

Ignoring the negative sign, the calculated value of EDI for the Manal mine is 0.41705.

To evaluate and verify the calculated EDI, three other mines, namely Gagal, Lambidhar and Kashlog limestone mines, which have almost identical geological setting but different operational phases, were selected, and the EDI was computed in the same way as was done for the Manal mine (Table 5.6).

From the computed EDI values, it can be inferred that at the Lambidhar mine the overall status of the environment is good. This mine was closed in 1996, and post-mining reclamation work has been completed. At the Kashlog mine, the overall environmental quality is worse than that at the Gagal mine. At the Manal mine, more stringent measures are required to improve the environmental status. Manal and Gagal mines are fully developed mines and production is in full swing. In contrast, the Kashlog mine is a comparatively new mine (operational since 1993). These computed EDI values for all four mines when cross-checked by ground realities and were found to be correct.

5.1.4 EDI versus Cost of EPC

Financial allocation is made in every mine for environmental protection. Here, we refer to this cost as *the EPC* which includes the costs of reclamation, planting of trees, green belt development, building of check dams and bunds and dust suppression, all of which are incurred for the conservation and protection of the environment.

Secondary data on investments per year towards environmental protection were obtained from the respective mines during field investigations (Table 5.7). A plot of the EDI versus the annual EPC per million ton of limestone produced was drawn, which is shown in Figure 5.7. In this figure, the line *AB* drawn at an angle of 45° shows the most ideal relation between EPC and degradation (in terms of EDI). If the intersection point of EPC and EDI falls on this line *AB*, one can infer that the company has invested sufficiently towards environment protection and preservation, and if this intersection is below or above this line, the EPC is either more or less than what it should be.

In a mine, if the environment degradation is more, the annual cost EPC will also be more and vice versa. The following important inferences can be drawn when the EPC and EDI for all four mines are plotted as in Figure 5.7:

1. At the Gagal mine, the investment towards environment protection (per million tons of limestone produced) is low. Similar is the situation at the Manal and Lambidhar sites.

   Interpretation: From *G*, draw the dotted lines *GG'G"* and connect *G'Y"* and *G'Y'* as shown in Figure 5.7. The line *G'Y'* represents the optimum cost required to be invested if the EDI is 0.2149. In the present case, since point *G* lies below line *AB*, *Y'* is also below *Y"*. Therefore, *Y'Y"* will be the extra cost required to match the optimum (ideal) cost of environment protection. Hence, more investment is needed to protect the sensitive environment in and around the mining area of the Gagal mine.

TABLE 5.7

Environmental Degradation Index and Environmental Protection Cost

Mine Name	Calculated EDI	Total Investment towards Environmental Protection (in Million Rs.)	Planned Production (MT)	Cost of Environment Protection per Million Ton of Limestone Produced
Manal	0.4170	0.4	2.8	0.142
Gagal	0.2149	1.2	3.0	0.4
Kashlog	0.2824	1.5	2.4	0.625
Lambidhar	0.1445	0.6	4.5	0.133

2. At the Kashlog mine, the EPC is above the ideal curve *AB*. Therefore, EPC equal to *K'K"* can be saved. It does not mean that the company should curtail the amount invested to protect the environment; however, based on these research findings, it can be concluded that the EPC in Kashlog mine area is sufficient to preserve the ecology and habitat.

5.1.5 Inference Drawn

Thus, it is evident that quantification of the affected environment by an industrial activity such as extraction of minerals on a numerical scale of 0–1 is possible, as explained in the previous section. It is also evident from the foregoing discussions that the decision-making process on whether the environment of any particular site is good or not can be made more definite, simpler and authentic. The adequacy of the investment by an organization towards environmental protection can also be confirmed scientifically. In order to make the EDI a powerful tool for environmental evaluation of a project, some scope still exists, as given here. Hence, further research is suggested and may be undertaken by others in the future.

1. Several new and modern statistical methods (computer based) have been perfected overtime. In future, other refined statistical calculation procedures that represent the overall impact values more closely and portray minimum error can be used to calculate the AQIN or WQIN index variables $X_1, X_2, X_3, X_4, \ldots$.

2. In the case of hill areas, noise and ground vibrations are important parameters to be considered for a complete EDI. Inclusion of these two parameters is necessary because of the fact that the topography of the hill areas is such that the noise produced gives rise to echoes and the noise impact becomes more pronounced. Similarly, ground vibrations get reflected from the lofty hill structures, with more severe impacts.

3. The water pollution parameters (e.g. pH, TSS and TDS) have different weightage in terms of the overall impact for a metalliferous mine, non-metalliferous mine and coal/fuel mine. Some parameters are not at all significant for some types of mined minerals. In this EDI development study, limestone is the mineral considered, and it was assumed that all parameters had equal weightage. This may not be true with all minerals; for example, a lead/zinc or copper mine may have more importance of the metal content in water. This area of 'weightage and importance to different pollution parameters' offers plenty of scope for future research.

4. For AQIN calculation, PM_{10} or $PM_{2.5}$ (in place of SPM) is the recommended air quality parameter set by various statuary agencies (Appendix 5.A). Their measured field values should be taken.

5.2 Part II: Environmentally Friendly Transportation Practices for Hilly Mines

Mining operation accentuates the twin problems of pollution (air and water) and land degradation. The majority of hill mines use road transportation with haul trucks/dumpers because of the flexibility offered by such systems. Such conventional modes of road transportation for minerals and overburden handling generate a huge amount of dust, which becomes a major source of air pollution. From the economics angle, also the running cost of dumpers in hills is higher than in the plains.

Small-scale mines make use of ordinary trucks (four-wheel drive) for bulk ore transportation in the hills. They have to, most often, take longer road routes for transportation from the hilltop to the bottom or from one hill to another. To avoid this circuitous path of hill roads and control the ill effects of air pollution, this section focuses on alternatives for bulk transportation of ROM/overburden that is being carried out in the surface mines or the hill quarries of the Himalayas. Other hill areas having similar undulating topography can also make use of these environmentally friendly systems.

On the basis of case studies, such transportation alternatives for bulk handling can be applied because there exists tremendous potential for their application. These techniques also offer some notable advantages; for example, they are easy to use, easy to install and economical in construction and can cut down the overall cost of mineral production. In Figure 5.8, the various alternatives to conventional road transport systems are analyzed and presented as 'Alternate A' to 'Alternate F'.

5.2.1 SFRC Chute (Alternate A)

Minerals can be transported from the hilltop to the bottom by an inclined chute (Figure 5.8) built along the hill side on the surface. The entire inclined chute, if necessary, can be supported on suitable structures and made in segments of varying length according to the requirement. The incline of the chute can be adjusted for maintaining a proper gradient. The lower half of the chute is made up of steel fibre–reinforced concrete (SFRC) to take care of the wear and tear due to the sliding of the mineral. The upper part of the chute can be made in such a way that it is easily detachable for cleaning the chute in case of necessity.

5.2.2 Vertical Silo with Conveyor Arrangements (Alternate B)

For an open-cast mine of 1 M tpa or more, the daily handling of ROM is on the order of 3500 tpd. In such cases, transportation by cylindrical bins and conveyor belts are the preferred alternative. The cylindrical bins of steel, also

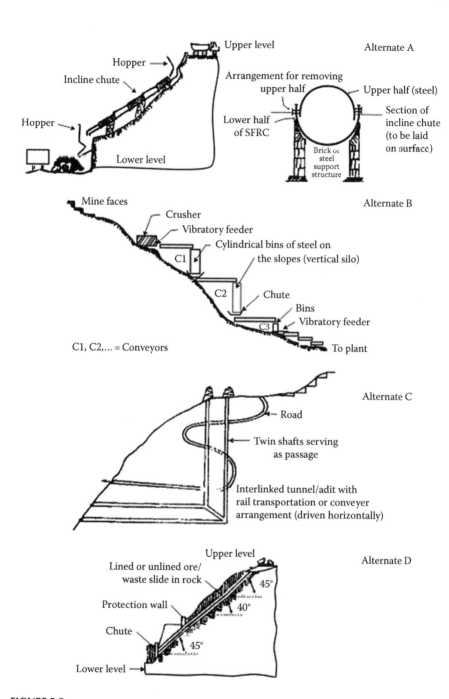

FIGURE 5.8
Environmentally friendly transportation practices for mines in hilly areas.

(*Continued*)

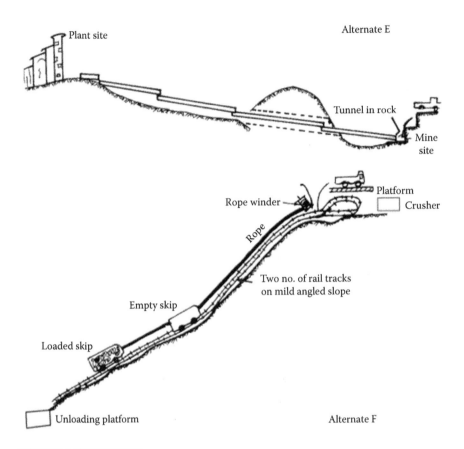

FIGURE 5.8 (CONTINUED)
Environmentally friendly transportation practices for mines in hilly areas.

termed as *surface silos*, act as a hollow shaft and can be erected vertically on the slopes by providing steel support frames.

Figure 5.8 shows a schematic representation of the cylindrical bin and conveyor arrangement. The layout is suitable for moderate to high production requirement. Vertical silos are made of thick steel sheets. The crusher location can be decided depending on its suitability and space availability. It can be located either at the hilltop or at the bottom. The whole transportation distance is covered in phases by the conveyors C1, C2, C3, C4, C5, etc. Arrangements for chutes, feeders, etc., can be done according to the design of the installation.

5.2.3 Shaft–Adit Combination or the SAC Method (Alternate C)

Mining of hill deposits usually commences from the hilltop, and to reach the top of the hill, an approach road, called a 'service road', is developed as a first step. Conventionally, this service road serves the purpose of transporting men and materials throughout the life of the mine and also fulfils the purpose of mineral transportation by dumpers. In the SAC method, a combination of shaft and adit, driven underground, is adopted in place of the service road for mineral transportation. From the mine benches to the top of the shaft, the usual rail or road transport system can be applied, as the shaft has to be located at a fixed location according to the mine lease, which is not frequently disturbed during the mine's life. From the top of the hill, a shaft can be driven to reach the lower levels, from where an adit may be excavated to connect the mineral body at the hilltop with the hill bottom. Construction of benches (mineral-producing faces) is done in the vicinity of shaft and radiating from it. The layout of internal roads connecting the bench with the shaft top can be drawn depending on the elevation, gradient and system of transport from the face to the shaft.

The shaft can be located at the centre of the property for ease in material handling. An alternate location of the shaft may be at the mineral boundary or at a lower horizon/level that the mining activity will reach after a certain period of excavation. Preferably, the shaft is to be driven in the host rock. The shaft top location should also be correlated with the length of the adit to be driven underground. The shape of the shaft and adit can be chosen as circular and horseshoe shaped, respectively, whereas their size depends on the other factors such as the quantity of ore to be handled, life of the project and economic considerations. A simple flowchart for the SAC method is shown in Figure 5.9. The shaft–adit combination has following two variants:

Variant 1: Transportation of ROM through shaft and crushing at the hill bottom

Variant 2: Transportation of the ore after crushing at the mining level (in-pit crushing)

The phases of ROM transportation in the SAC method are as follows (Figure 5.10).

Phase I: Transportation from the bench face to transfer point No. 1, that is the shaft mouth

Phase II: Transportation from transfer point No. 1 to transfer point No. 2, that is at the shaft bottom

Phase III: Transportation from the shaft bottom to the transfer point No. 3, that is the crusher, from where the mineral is sent to the plant or siding (as the case may be) by belt conveyors

Phase IV: Transportation from transfer point No. 3 to transfer point No. 4, that is from the crusher to the conveyor belt

5.2.4 Ore Slides in Rock (Alternate D)

Ore slides are inclined passages built in the rock itself and slightly differ from the inclined chute. The installation is built either by 'cut and cover' or

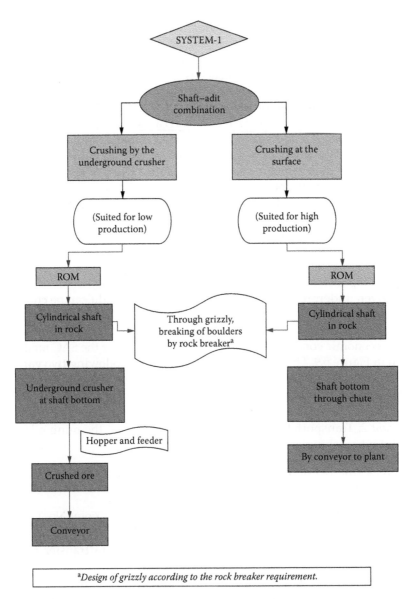

FIGURE 5.9
Flowchart of the SAC concept.

by underground construction techniques, and as such no supporting structure is visible on the surface. The inclination of ore slides is decided in such
a way that the material/mineral flow from hilltop to hill bottom is smooth
and unhindered. Such ore slides are provided with an upper-level unloading platform for dumpers/trucks and chutes at lower level for drawing the

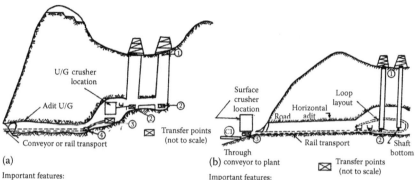

FIGURE 5.10
Schematic representation of different variants for (a, b) shaft–tunnel combination as an alternate for ore transportation. (a) alternate No. 01 and (b) alternate No. 02.

mineral/ore. The ore slide can be either lined or unlined and provided with arrangements for taking care of problems such as chocking/jamming.

Chocking/jamming of the chute/slide is one big practical problem that may be encountered in limestone mining because limestone is a sedimentary rock and its association with clay is quite common. The clay, being sticky in nature, creates a problem especially during rainy season by the agglutination of smaller chunks to form a big one, thereby clogging the restricted passage in a chute.

5.2.5 Overland Belt Conveyor (OLBC, Alternate E)

Conveying the raw material using a conveyor laid down on the surface is quite common in practice in mines and plants handling bulk solids. As the name suggests, OLBC is a conveyor system laid on a surface. The only difficulty in a hill topography is the gradient of the conveyor which can be easily controlled by its alignment in an appropriate manner. For this purpose, construction of some underground structures on the rock/ore, such as a tunnel, is required, if necessary. Sometimes, OLBC is also called a 'cross-country belt conveyor' system.

5.2.6 Gravity-Operated Skips on Guided Rails at Surface (GOSS, Alternate F)

This transportation alternative takes advantage of gravity for the movement of skips on guided rails which are laid on the surface. Such a system requires virtually no power for the skip movement and can be easily fabricated at local workshops. GOSS has limitation for operation on a steep slope, that is more than 50°, because of control and safety reasons. Ropes required for the attachment must be strong enough. Proper safety attachments with skips are a must with such installations. They are cost-effective and environmental friendly alternatives for both small- and medium-scale mines.

5.2.7 Other Advanced Alternatives

Advancement of technology has led to some costly yet environmentally friendly means of transporting minerals over long distances in the hills; but they are operational only in some western countries and not in India. They are briefly described here to complete the description of environmentally friendly alternatives.

Slurry transportation: Transportation of the excavated material is also possible by the well-established method of slurry transportation. This can be done by mixing the raw material with water in the desired proportion and pumping the slurry through pipelines (Basu and Saxena, 1984). At the receiving terminal, the slurry is dewatered and/or dried by a process for the use of the mineral. Certain experiments need to be carried out, and water should be available in large quantities for assessing its feasibility. In a hilly terrain, gravity can be taken as an advantage for its implementation.

Pneumatic capsule pipeline system (PCPS): PCPS is another advanced pipeline capsule system for the transportation of minerals, and it was first introduced in former Soviet Union three decades ago. This system has been economically operating since 1983 for the transportation of bulk solids from the mine to the plant at the Karasawa mine in Japan (Kennedy, 1993). The basic principle for PCPS is as follows: load a capsule with the mineral, blow or suck the capsule along a pipeline and unload it at the other end. The system is useful for uphill transportation (at a high gradient of more than 45°) where spillage of the transported ore is a major issue. A schematic diagram showing the main features of the capsule liner transportation system is given in Figure 5.11.

In order to assess the possibility of ROM transportation of mineral in a mine by various alternatives, two types of feasibility study need to be conducted: (1) evaluation of existing network/system and (2) techno-economic feasibility of the proposed alternatives (Kutschera, 1984). Their economic viability and suitability for the site is a subject matter of detailed investigation to be carried out on a site-specific or an individual basis.

Some of the transport alternatives described earlier are suitable for small-scale production, whereas some can satisfy medium and large requirements.

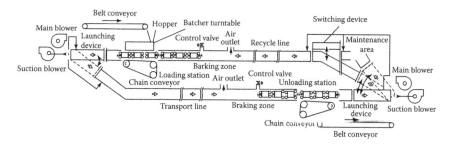

FIGURE 5.11
Pneumatic capsule pipeline system for the transportation at the Karasawa Limestone Mine, Japan: main features.

Their economics will be the driving force for selection and implementation. Hence, it is quite clear that efficient transportation makes a mine more profitable, and the overall cost of production can be brought down substantially in the long run. This way, ecologically fragile areas, which suffer from adverse impacts and negative imbalances of pollution, can be kept safe and productive at the same time.

Appendix 5.A: Air Quality and Water Quality: Prescribed Indian Standards

5.A.1 Air Quality

Air pollution in open-cast mining areas is mainly caused by various unit operations such as drilling/blasting, excavation, crushing and transportation. During mining and mine operation, dust and particulate matter are emitted into the air. The haul road is a perennial source of dust, which when becomes airborne, adversely affects the productivity and the working environment. Thus, the presence of these particulate matters of various sizes (including dust) in mining areas is a serious problem for the management and is also a health hazard for the engaged workforce.

For air pollution in the mining areas, suspended particulate matter (SPM), respirable particulate matter (RSPM) and particulate matter (PM_{10} and $PM_{2.5}$) are the significant parameters along with sulphur dioxide (SO_2), carbon monoxide (CO) and other noxious gases (NO_x).

Previously, SPM and RSPM were the major pollutant *parameters* in the case of ambient air quality in the mining environment. But later research showed that particulate matter of less than 10 µm size is problematic in particular for human health, whereas oversized particles (i.e. >10 µm size) are easily managed by human body. This is the reason why the parameters PM_{10} and $PM_{2.5}$ are considered for air quality assessment these days. Accordingly, the national statuary bodies (CPCB for India) prescribe the various limits for these parameters (Table 5.A.1).

The SPM, RSPM, SO_2 and NOx concentration should be measured in the downwind direction considering predominant wind direction at a distance of 500 m from the dust-generating sources and should not exceed the prescribed standards specified by the national statuary agency. The dust-generating

TABLE 5.A.1

Air Quality Parameters for AQIN Determination

| Parameter (in μg/m³) | Concentration in Ambient Air | | | | | |
| | Industrial Areas | | Residential Areas | | Sensitive Areas | |
	Upper Limit (UL)	Lower Limit (LL)	Upper Limit (UL)	Lower Limit (LL)	Upper Limit (UL)	Lower Limit (LL)
SPM*	430	600	360	500	360	500
RSPM* (size < 10 μm)	250	300	215	300	180	250
PM_{10}	60	100	60	100	60	100
$PM_{2.5}$	40	60	40	60	40	60
SO_2	50	80	50	80	20	80
NO_x	40	80	40	80	30	80
CO	02	04	02	04	02	04
Pb	0.50	1.0	0.50	1.0	0.50	1.0

Source: CPCB, National ambient air quality standards, 2009. http://cpcb.nic.in/National_Ambient_Air_Quality_Standards.php.

Notes: 1. Methods and the duration of air quality measurement should be as per the CPCB India Notification dated 18 November 2009.

2. Here, the asterisk (*) means that these parameters are considered by CPCB for coal mines of India as industry-specific standards under rule 2(1) of the Environment (Protection) Amendment Rules, 2000, notified vide notification G.S.R. 742(E), dated 25.9.2000, applicable for India under the Environment (Protection) Rules, 1986; http://cpcb.nic.in/Industry-Specific-Standards/Effluent/494-1.pdf.

TABLE 5.A.2

Air Quality Standards as Prescribed by the Indian Standards Organization

S. No.	Category	SPM	SO_2	CO	NO_x	Respirable Dust Concentration
1.	Environmentally sensitive areas	100	30	1000	30	75
2.	Industrial areas and mixed uses	500	120	5000	120	150
3.	Residential and rural areas	200	80	2000	80	100

Source: CPCB, New Delhi, India.
Note: All values are in μg/m³.

areas are the loading/unloading of ROM; haul roads; approach road; transportation all along the mine, that is haul road and approach road; ore/coal handling plant (CHP), blasting and drilling operations, overburden dumps or any other dust-generating external sources.

It may be noted that PM involves two subcontributions, namely PM related to the main activity, that is mining (which includes all related ancillary operations except vehicle exhaust), and PM related to road transport activity that is indirectly released from the vehicle exhaust. Detailed research on non-exhaust PM emission (Pandian and Prasad, 2014) and emission factor determination (Ghose, 2004) for different mining areas are being carried out in India and abroad.

For the Himalayas, sensitive areas values, as given in Table 5.A.2, should be considered.

5.A.2 Water Quality

Similarly, for water quality assessment, there are 35 constituent parameters such as pH, TSs, DSs, BOD and COD (Table 5.A.3) which are examined to know the level of pollution.

TABLE 5.A.3

Water Quality Standards for Discharge of Effluents (per Schedule I, Rule 3 of Environment Protection Rules, 1986)

		Standards			
		Inland Surface Water	Public Sewers	Land for Irrigation	Marine Coastal Areas
S. No.	Parameters	1	2	3	4
1.	Colour and odour	See Note 1	–	See Note 1	See Note 1
2.	Suspended solids (in mg/L max.)	100	600	200	(a) For process waste water (100)
					(b) For cooling water effluent = 10% above total suspended matter of effluent cooling water
3.	Particle size of suspended solids	Should pass 850 μm	–	–	(a) Floatable solids (mg/L max. IS sieve)

(Continued)

TABLE 5.A.3 (*Continued*)

Water Quality Standards for Discharge of Effluents (per Schedule I, Rule 3 of Environment Protection Rules, 1986)

		Standards			
		Inland Surface Water	Public Sewers	Land for Irrigation	Marine Coastal Areas
S. No.	Parameters	1	2	3	4
4.	Dissolved solids (inorganic, mg/L max.)	2100	2100	2100	–
5.	pH value	5.5–9.0	5.5–9.0	5.5–9.0	5.5–9.0
6.	Temperature (in °C)	Should not exceed 40 in.; the section point of the stream within 15 m downstream from the effluent outlet	45	–	45 at the point of discharge
7.	Oil and grease (mg/L max.)	10	20	10	20
8.	Total residual chlorine (mg/L max.)	1.0	–	–	1.0
9.	Ammoniacal nitrogen (as N) (mg/L max.)	50	50	–	50
10.	Total Kjeldahl nitrogen (as N) (mg/L max.)	100	–	–	100
11.	Free ammonia (as NH_3) (mg/L max.)	5.0	–	–	5.0
12.	Biochemical oxygen demand (5 days at 20°C) max. (mg/L max.)	30	350	100	100
13.	Chemical oxygen demand (mg/L max.)	250	–	–	250
14.	Arsenic (as As) (mg/L max.)	0.2	0.2	0.2	0.2
15.	Mercury (as Hg) (mg/L max.)	0.01	0.01	–	0.01
16.	Lead (as Pb) (mg/L max.)	0.1	1.0	–	1.0

(*Continued*)

TABLE 5.A.3 (*Continued*)

Water Quality Standards for Discharge of Effluents (per Schedule I, Rule 3 of Environment Protection Rules, 1986)

S. No.	Parameters	Standards			
		Inland Surface Water	Public Sewers	Land for Irrigation	Marine Coastal Areas
		1	2	3	4
17.	Cadmium (as Cd) (mg/L max.)	2.0	1.0	–	2.0
18.	Hexavalent chromium (as Cr^{6+}) (mg/L max.)	0.1	2.0	–	1.0
19.	Total chromium (as Cr) (mg/L max.)	2.0	2.0	–	2.0
20.	Copper (as Cu) (mg/L max.)	3.0	3.0	–	3.0
21.	Zinc (as Zn) (mg/L max.)	5.0	15	–	15.0
22.	Selenium (as Se) (mg/L max.)	0.05	0.05	–	0.05
23.	Nickel (as Ni) (mg/L max.)	3.0	3.0	–	5.0
24.	Boron (as B) (mg/L max.)	2.0	2.0	2.0	–
25.	Percent sodium max.	–	60	60	–
26.	Residual sodium carbonate (mg/L max.)	–	–	5.0	–
27.	Cyanide (as CN) (mg/L max.)	0.2	2.0	0.2	0.2
28.	Chloride (as Cl) (mg/L max.)	1000	1000	6000	–
29.	Fluoride (as F) (mg/L max.)	2.0	15	–	15
30.	Dissolved phosphate (as P) (mg/L max.)	5.0	–	–	–
31.	Sulphate (as SO_4) (mg/L max.)	1000	1000	1000	–
32.	Sulphide phosphate (as P) (mg/L max.)	2.0	–	–	5.0
33.	Pesticides	Absent	Absent	Absent	Absent

(*Continued*)

TABLE 5.A.3 (*Continued*)

Water Quality Standards for Discharge of Effluents (per Schedule I, Rule 3 of Environment Protection Rules, 1986)

		Standards			
		Inland Surface Water	Public Sewers	Land for Irrigation	Marine Coastal Areas
S. No.	Parameters	1	2	3	4
34.	Phenolic compounds (as C_5H_6OH) (mg/L max.)	1.0	5.0	–	5.0
35.	Radioactive materials				
	(a) Alpha emitters ($\mu C/mL$ max.)	10^{-7}	10^{-8}	10^{-8}	10^{-7}
	(b) Beta emitters ($\mu C/mL$ max.)	10^{-6}	10^{-5}	10^{-7}	10^{-6}

Notes: 1. All efforts should be made to remove colour and unpleasant odour as far as practicable.
2. The standards mentioned in this notification should apply to all the effluents discharged such as industrial mining and mineral processing activities and municipal sewage.
3. This notification will not apply to those industries for which standards have been notified by the central government vide S.O. 844(E) dated 18 November 1986, S.O. 393(E) dated 16 April 1987, S.O. 443(E) dated 28 April 1987 and S.O. 64(E) dated 18 January 1988. This notification will cease to apply with regard to a particular industry when industry-specific standards are notified for that industry.
4. Schedule II is added by notification no. GSR 919(E) dated 9 December 1988 published in the Gazette of India, Extra, Part II Sec. 3(i) dated 9 December 1988.

For air and water quality assessment, some parameters have a single prescribed limit and some have a range of values (e.g. pH values and dissolved solid values of water quality). In case no lower limit is prescribed, a value of 0 can be taken and the single prescribed value can be considered as the upper limit for AQIN or WQIN determination.

6

Integrated Strategy and Best Practice Mining

The Himalayas, a classic case of geological complexity and environment fragility, were created by continent–continent collision. The research on the Himalayas is very vast and rich, spanning over 150 years, making the literature almost intractable (Yin, 2006). A number of researchers have studied the Himalayas from the perspective of their discipline, which in turn has accelerated the rate of publication on more and more specialized subjects. However, the subject area of Himalayan mining has remained under explored. In the past two decades, the scientific mining of minerals in the Indian Himalayas has been considered to some extent, and some good work has been done. While compiling and citing these works in this book, we realized that the negative impacts of mining on the various components of the environment should be dealt with sensibly. Environmental disturbances due to mining must be controlled and contained in the localized areas only using environmentally friendly methodologies. The adopted practices should be such that mining and environmental protection can go together in a sustainable manner.

Keeping this in view, the problems associated with Himalayan mining are dealt with in this chapter, considering the inherent features of the region and likely damages that mining might cause to the ecosystem. A descriptive attempt has been made in two parts: (1) an integrated development planning and approach for the resource management of Himalayan mineral-bearing areas and (2) best practice mining involving its problems and solutions. This has been done by amalgamating large-scale and small-scale mining that are going on in the Indian Himalayas.

6.1 Integrated Approach/Strategy

Integrated strategy (IS) can be defined as a 'scientific approach to rational mountain development which causes least harmful environmental effects of any developmental activity whether mining or otherwise and is a multidisciplinary strategy involving the co-operation and collaboration of local population with experts and decision makers thereby identifying and bridging the gaps in the knowledge with the ultimate objective of fulfilling the aspirations of local populace'. Hence, whenever an 'approach' or 'strategy' is referred to in the context of the Himalayas, the word *integrated* should be used along with it to give it a holistic view. Important attributes of an integrated approach are as follows:

- It meets the requirements for maintaining biological productivity.
- It safeguards the ecological balance and ensures rational development of the mountain areas.
- It calls for the optimal use of the mountain resources, which can be sustained over several generations in the context of available technology.
- It reconciles the economic needs and aspirations of the local inhabitants.

In order to facilitate area-specific microlevel planning for the management of areas with mineral-bearing resources, it is best to apply an integrated approach on a watershed basis. Generally, the term *watershed* refers to a catchment or drainage basin and is used in relation to water only. But *a watershed and its management* are not only related to water; it is the rational utilization of land resources and water resources for optimum production with minimum hazard to the natural resources. It is essentially related to resource conservation, which means proper land use, protecting the land against all forms of deterioration, building and maintaining soil fertility, reducing sediment load, conserving water, controlling water drainage, properly managing food and fodder and increasing productivity from all land uses.

Watershed, as a unit of environmental planning and management, constitutes a technical and ecological entity that is self-contained, physically composite and functionally coherent for land, water and air resource planning and management (Gupta, 1983). It is compatible with the existing block- or village-level planning. According to the multilevel planning policy of the Government of India, at the national, state, district and lower levels, natural resource data management is done on a watershed basis, considering each watershed as a constituent unit for planning. Hence, in ecologically fragile areas of a hilly region containing mineral resources of economic importance, the study of environmental parameters and other related factors should be done on a watershed basis. Integrated strategy on a watershed basis for land utilisation, soil and water conservation as well as utilization or recycled waste can be rationally articulated for comprehensive short-term and long-term planning and their effective implementation. This approach provides resource-centred production and is environmentally benign, thereby helping in promoting sustainable development.

Watershed-based integrated planning is natural endowment based and hence varies according to the shape, size, relief, soil type, land use, hydrology and vegetative cover. According to the sizes of watershed in which the mine area falls, macro, micro or mini watershed planning can be done

(Dhurvanarayan et al., 1990). If a watershed has an area between 5000 and 3000 ha, it is called a 'macro watershed' and if the area is in between 3000 and 1000 ha, it is called a 'micro watershed'. If the land area is of still smaller dimensions, namely less than 1000 ha, it is called a 'mini watershed'. Apparently, the classification of a watershed is rather loose. For an 'agriculture-dominated land' and for a 'river basin', the sizes may be different, but the basis of deciding the size of watersheds is only the drainage pattern, that is an area of land that drains into a common outlet.

Similarly, another method of watershed planning is based on drainages as *first*, *second*, *third* or *higher* order. A first-order watershed is one in which the drainage is from the watershed into a single, unbranched stream channel, whereas in a second-order watershed two first-order channels come together to form a second-order channel and so on.

For optimum conservation of water, land (soil) and minerals, due regard should be given to advance planning and management with an interdisciplinary, scientific approach, as described in the chapter.

6.1.1 Developmental Planning

Owing to rapid industrialization, environmental problems in the mineral sector are reaching a crisis point. Although the government has been trying to solve these problems by various means, the main emphasis has been on end-of-pipe solutions. This section describes a strategy for development that suits the selected sensitive Himalayan area. The core environmental areas, namely water, air, land (deforestation and waste handling), ecology and society (energy utilization and quality of life), are discussed with regard to its implementation. An empirical approach for developmental planning has been formulated and elaborately described with an aim to harness its benefits by the user industrial organizations.

Though the term *development planning* encompasses a wide area with no yardsticks and ends, broadly it is an approach which includes the major environmental management parameters. Table 6.1 gives the details of the environmental parameters according to the valuation or importance, which is to be taken into account for a holistic approach to the development strategy. The strategy must be an 'integrated' one to decide the future course of action. The weightage is assigned to all the six major environmental parameters, according to their importance, on a scale 0–100. The most important component in the Himalayan region is ecology, and thus, it has been assigned 30% weightage. The societal component of the environment and land has been assigned equal weightage of 20% each, as all the human activities are for the benefit of the society and land based. Water, air and noise have been given 10% each, as the condition of ecology, society and land depends to some extent on these parameters. Thus, the total environmental quality score of 100 can be categorized as follows for the mining areas of the Himalayas.

TABLE 6.1

Environmental Parameters and Their Weightage

S. No.	Environmental Parameters	Components	Weightage	Remarks
1.	Ecology	Flora and fauna	30	Encompasses all types of biodiversity of vegetation cover and wildlife
2.	Society	Quality of life (QOL)/total environmental quality (TEQ)	20	QOL/TEQ includes the following important parameters/indicators Housing Water facility Food and nutrition Health and medical facilities Education Income per family Communication and transport Assets Recreational facilities
3.	Land	Topography, drainage, land use, land form	20	Extremely important for hills, as they are short in supply. Soil is a scarce commodity
4.	Water	All 35 water quality parameters	10	Selection of parameters and their importance according to the end use of water
5.	Air	$SPM/PM_{10}/PM_{2.5}$, SO_2, NO_x, CO, CO_2, Pb, hydrocarbons	10	Workplace air quality parameters are important compared to ambient air quality.
6.	Noise	Road/rail transport, blasting	10	Caused by traffic movement and unit operations of mining (blasting leads to ground vibration and fly rocks a causative factor for noise)

Note: The parameters are arranged in descending order of their importance and applicable for the Himalayas (fragile hill areas).

Scale	Environmental Quality
0–20	Very poor
20–40	Poor
40–60	Fair
60–80	Good
80–100	Excellent

6.1.1.1 Ecology

In the Himalayan context, ecology is the most important parameter of the overall environment because of the climatic factors, biodiversity and changes in ecological composition with altitude. The diversity at the same altitude with changed locations marks the characteristic features of fragile hill areas. The vegetation and its density, hill slope including their stability and soil erosion due to rain and wind versus human interference have impacts on this component.

Figure 6.1a shows the exponential population trend with time. In Figure 6.1b, the ecological parameter versus time for growth and natural extinction are shown by two separate lines. For a particular time, the quantum of flora, fauna and wildlife in an ecosystem can be worked out as $R = X-Y$. The trend

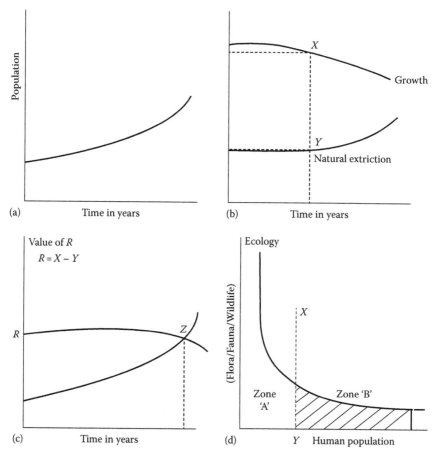

FIGURE 6.1
(a–d) Ecological component of integrated strategy.

of R when drawn with the population time curve (Figure 6.1c) gives the point Z as the limiting point. Figure 6.1d explains that, as the human population increases, the density of flora, fauna and wildlife decreases. This exponential downward trend has a cut-off line (X–Y) dividing the ecology versus population curve into two zones, 'A' and 'B'. Any ecosystem has got its sustenance capacity (resistance/resilience) for a limited amount of population, and zone 'A' is the safe one from the ecological safety point of view. Thus, developmental plan should be such that the coordinates depicting the ecological parameter with respect to population should lie in zone 'A'.

This trend holds good for both flora (biological life and vegetation) and fauna (animals and wildlife).

6.1.1.2 Society

All the human activities in the various regions, including the Himalayas, are for the development of the society, and these activities primarily depend on the various resources available. The societal parameters in relation to environmental quality can be comprehensively described by the term QOL for the quality of life (Prushti, 1996; Saxena et al., 2002). However, QOL is a broad term, and no consensus has emerged on an accurate definition of the term and its uniform applicability for the obvious reason that it varies with time and from one geographical domain to another. The following essential indicators (parameters) can be used to describe the QOL:

- Shelter (home/house)
- Water facility
- Food and nutrition
- Education
- Income per family and assets
- Health and medical facility
- Communication and transport
- Recreation facility
- Fuel and energy
- Sanitation

Poor living conditions tend to put more stress on the ecosystem, and reasonable/good/better quality of life gives less stress. 'Value function curves' for each of the 10 parameters can be drawn for specific locations or regions of the Himalayas to define QOL. For the present purpose, these indicators are divided into four groups (I–IV), and their variation with population pressure is depicted in Figure 6.2.

• Shelter	Basic amenities	Group I
• Water		
• Food		
• Health and medical facility	Infrastructural facility	Group II
• Fuel and energy		
• Communication and transport		
• Recreation facilities		Group III
• Sanitation		
• Education		
• Income and asset per family		Group IV

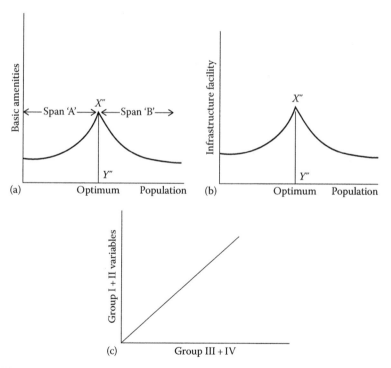

FIGURE 6.2
(a–c) Societal component of integrated strategy.

By assigning equal weightage to all the parameters, the QOL of each of the parameters can be defined in the range 0 (worst quality) to 1 (best quality). The rating of all the parameters is then added to get a single numerical value, which is referred here as the *quality of life index* (QOLI). For QOLI of the individual families in a mining complex (from 1 to 10), the following grades are suggested.

QOLI	Grade
<2	Very poor
2–4	Poor
4–6	Fair
6–8	Good
8–10	Very good

Considering Figure 6.2a and b separately, it is seen that a similar behaviour of human population is reflected on the basic amenities and infrastructural facility curves. Thus, it can be said that up to a limited population level (line X'–Y' and X''–Y''), both basic amenities and infrastructural facility are sufficient for maintaining a reasonable QOL. If span 'A' is shorter than span 'B', human population pressure on the basic amenities and infrastructures is limited. Group I and II variables are balanced if they lie in span 'A'. Qualitatively, group III and IV parameters are directly proportional to group I and II variables. It means that if the basic amenities and infrastructural facilities are satisfactory, education, income and asset will also be satisfactory (Figure 6.2c). Based on this, the QOL for the Himalayan area on a scale of 1–10 can be worked out, and development strategies can be planned accordingly (Soni, 1997).

This discussion has also brought out that the societal parameters influencing the environmental imperatives and mining practices, which are very vast, leave much scope for interpretation depending on their personal understanding. Hence, areas of inadequacy in environmental management can be identified on a case-to-case basis, and suitable remedial measures can be adopted by the management of mines during planning and design.

The other major environmental management areas, namely land, water, air and noise, for the developmental planning can be effectively dealt with thorough 'best practice mining', which is an engineering-based approach, as described later in this chapter. Numerous ways and means are available for these major core areas to remedy the identified lacunas.

On a site-to-site basis, an empirical approach, an analytical approach or a numerical analysis approach can be used to manage the environmental quality.

6.1.2 Benefits of Advanced Development Planning

It is an acknowledged fact that roughly 1/10 of the world population lives in the hilly regions (Tables 6.2 and 6.3) and nearly 40% directly or indirectly depend on mountain resources, that is forests, water, agriculture, minerals or tourism. In India, the hill population comprises about 9% of the total population. The hill area is about one-fifth of the total land area. The economy

TABLE 6.2

Distribution of Population in the Alps

Country	Percentage of Land Covered by the Alps	Percentage of the Population That Lives in the Alpine	Alpine Area (100 km²)	Population Density in the Alpine Region (inhabitants/km)
Italy	16	6	48.7	71
Austria	63	40	48.3	61
France	7	4	35.1	61
Switzerland	58	18	24.6	74
Yugoslavia	7	7	18.2	78
W. Germany	2	1	4.9	102

TABLE 6.3

Distribution of Population in the Himalayas

Country	Percentage of Land Area Covered by Hilly Himalayas	Percentage of Population That Lives in Hill Areas	Remarks
India	20	9	Besides the Himalayas, other hill areas also exist
Nepal	72.9	62	–
Bhutan	100	100	It includes some low-land areas as well that are not lofty hills.
China	6.66	3.33	–
Pakistan	NA	NA	–
Afghanistan	66	60	Afghanistan has three main land regions: North Plains, Central Highlands and SW lowlands. The Central Highlands cover two-thirds of Afghanistan and consist of the Hindukush Himalayas whose data are reproduced here. Northern plains of Afghanistan also contain some mountainous areas whose data are not available and hence not included.

of Nepal and Bhutan is largely dependent on Himalayan resources as they are the total hill countries, and all their human needs are derived from hill resources.

Fragile areas with a mountainous topography have dynamic characteristics owing to the sensitiveness and complex interrelationship between various environmentally related parameters. To search for the best solution keeping in mind the alert perception of public towards developmental projects, it is necessary that ecological and social parameters should be given highest importance for fragile areas. This will enable sustainable and scientific development of mineral-bearing areas. An integrated developmental planning takes into consideration all these important components and hence will be able to promote sustainable development and environmental management of a mining area in entirety.

6.2 Large-Scale versus Small-Scale Mining (SSM) in the Himalayas

Concentrated output can be achieved from large deposits. But since only selected large-scale mechanized mining is done in the Himalayas for limestone and magnesite, we have to look towards both small- and large-scale mining. In actual practice and compared to the organized large sector, the scenario of mining is different in the sense that SSM accounts for more than 50% of the total production in the Himalayas. Thus, the SSM sector, which is largely unorganized, contributes to the most production, and therefore, due consideration must be given to the artisanal or SSM sector.

As opposed to large-scale mines, small-scale mines are characterized by less requirement of reserves, implementation time, initial investment and high employment potential with moderate levels of skills and infrastructure requirements (Figure 6.3). It is observed that naturally developed small mines take longer time to develop and have drawbacks in terms of implementation

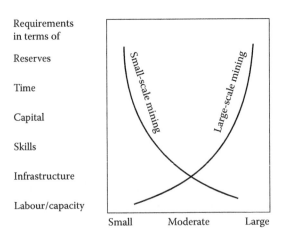

FIGURE 6.3
Comparative profile for small- and large-scale mines. (After Noetstaller, R., Small scale mining: Practices, policies, perspectives, in: Ghose, A.K., ed., *Small Scale Mining: A Global Overview*, Oxford IBH & Company, New Delhi, 1994, pp. 3–10.)

of scientific approaches, but well-planned and executed small mines, if developed with imagination and long-term vision, can shape the actual mining operation on modern lines in an eco-friendly manner.

Thus, broadly it can be argued that the artisanal/SSM sector will continue to employ conventional methods of mineral exploitation, and for this, low-cost and effective engineering solutions are required. To a larger extent, this can be done by 'best practice mining' and by the implementation of an 'integrated approach' for planning and execution. Various steps right from the planning stage to the reclamation stage should be taken in this regard.

6.3 Best Mining Practice

Best mining practice (BMP) or *best practice mining* (BPM) does not refer to any designed/formulated method but implies continuous improvement of existing practices so that the negative impacts of mining are minimal (EPA, 2002; Tripathy and Reddy, 2015; USEPA, 2001). To follow BPM, developmental planning requires some prior considerations, such as those given in the following:

- Watershed must be a unit of the environmental management in the Himalayas. Since the Himalayan region is characterized by sharply changing features going from one watershed to another, a watershed development management plan for the total life span of the project should be from an essential constituent of planning.

- While designing and implementing a mining project (as part of the developmental activity for the region), at both macro- and micro-planning levels, explicit considerations should be given to the characteristic features of the mountainous region, which are termed *mountain specificities* and include inaccessibility, fragility, marginality, diversity and niche (ICIMOD, 1983).

- Guidelines for the grant of mineral rights should be framed beforehand, keeping in view the hill area type and its delicacy. These guidelines should be drawn in consultation with technical experts and representatives of the mine owners so that they are practical in implementation.

- Small-scale mines, which account for a considerable percentage of the production, need special mention and attention. To implement BPM in small mines, due consideration for environmental and mining problems on a site-specific basis must be given. It includes the following:

  - Creation of green barriers along haulage roads and along the periphery of the outer limit of the quarry

- Projected parapet walls along the approach roads with necessary super elevations
- A mobile environmental monitoring unit for a group of mines and an internal wing of environmental management for various mining operations
- Community nursery
- Use of a hydroseeder for the reclamation of highly angled slopes
- Progressive restoration and scientific reclamation practices
- Use of rippers as an alternative to blasting
- Controlled blasting and use of sequential blasting machine to control fly rocks and ground vibrations
- Gravel/sand-packed pits to check water pollution
- Improved design of check walls/check dams with filtering arrangements
- Environmental-friendly and scientific methods of transportation for a cluster of small mines
- Environmental-friendly equipment, machine and accessories, like continuous miners
- Scientific methods of waste disposal by backfilling or by fill construction in lifts

In the hill areas, the concepts of plains are obviously not applicable and, thus, need to be restructured and inducted into the mining process right from the planning stage to the reclamation stage.

To best explain BMP for the Himalayas in the following paragraphs, the issues are divided into two parts: 'environmental problems' and 'mining problems'. Accordingly, for each problem, its solution(s) as per the integrated strategy are given.

ENVIRONMENTAL PROBLEMS
Problem 1
Lack of environmental protection measures

Solution

1. Tree-lined approach roads should be the preferred alternative in place of ordinary roads in small mines. Laying of such roads should be encouraged in all SSM areas (Figure 6.4). The importance of creation of tree barriers along the approach road (not the haulage road) and along the periphery of the outer limit of the quarry, that is mine lease boundary, at the initial stage of the mine development should

FIGURE 6.4
Tree-clad approach roads.

be insisted upon. If this is done, it will serve the following two purposes:

a. It acts as a barrier for dust, fines and flying fragments outside the mine premises. Trees combat dust and noise pollution in a natural way.

b. It acts as a curtain to cover the ugly scars of the degraded land created by road development activity.

2. A large amount of dust is generated during mining. This dust can be utilized for making agglomerates/pellets, which can be used as a raw material feed.

3. A mobile environmental monitoring laboratory should be commissioned for a group of small mines on cost-sharing basis (Figure 6.5).

4. To take up the responsibility of environmental management at the operational stage of the mine, an 'internal wing of environmental management of mining operation for a cluster of mines' is desirable with the integrated strategy, as this aspect is ignored in small mines.

5. Development of a 'community nursery' for tree planting by a group of small mines and during the preproduction period.

Problem 2

Lack of synthesis of social and industrial needs for environmental protection

Solution

The restoration of mined-out land should be done to put it to various end uses. This includes the development of the area for recreational activities, for example as playground, sports field, polo ground, pond for rainwater

FIGURE 6.5
Mobile environmental monitoring van suitable for a group of small-scale mines.

harvesting or pisciculture or for other civic uses. When the mined-out space is restored, it is possible for local habitants to settle in the area if basic amenities are provided. In this way, the greenery of the area can be restored.

Thus, the synthesis of the social and industrial need has the potential to cultivate a true eco-friendly culture towards mineral exploitation in the Himalayas.

Problem 3

Difficulties in reclaiming high-angled hill slopes in and around mining areas that are disturbed as a result of mining or other developmental activities

Solution

For this purpose, the *hydroseeder* could be made use of (Figure 6.6). This technique involves the spraying of a mixture of soil, organic matter, grass seeds, adhesives and water in a fixed proportion, which is kept in a slurry tank. The application of the mixture on the slope is done under pressure. A prototype model of the hydroseeder has been developed by the Forest Research Institute, Dehradun, India. The viability of this technique for the reclamation of mine dumps has been well tested in the limestone and rock phosphate mines in the Dehradun and Mussoorie regions.

Problem 4

Lack of restoration of excavated mining areas on scientific lines according to the approved environmental management plan (EMP)

FIGURE 6.6
'Hydroseeder' for growing vegetation on steep hilly slopes.

Solution

Scientific and environmental-friendly ways for restoration broadly include the following:

- Recovery from mining phase.
- Spreading of topsoil.
- Progressive restoration of land quality by biological reclamation in a permanent way. The integrated biotechnological approach (Box 6.1) is the best solution.

BOX 6.1   INTEGRATED BIOTECHNOLOGICAL APPROACH (IBA)

For Reclamation and Revegetation of Mine Spoil

The IBA is a method in which appropriate blending of spoil and organic waste(s) is done for the establishment of the plant microflora and inoculation of plants with specialized culture(s) of nitrogen-fixing microorganisms. Strains of endomycorrhizal fungi are used for profuse root development and stress tolerance in plants. The process promotes resource conservation through waste utilization and leads to fast recovery of degraded ecosystem (Figure 6.7). It provides carbon dioxide sinks, builds up fertile topsoil and generates fibre, food, fruit and fuel.

Pits of suitable sizes are excavated on the stony mine spoil, which include dumps as well as degraded stony land areas. In each pit, bedding material consisting of organic waste and spoil is filled. Pretreated saplings of select species of high economic/ecological value with select microbial cultures using the root inoculation technique are planted in the pits. The vesicular arbuscular mycorrhizal fungal spore inoculum is also added in the rhizosphere of the plants. No chemical additives are used in the process. Within a period of 3–4 years, the degraded land/dump surface is reclaimed, which looks green.

This methodology has been successfully used in manganese and coal spoil dumps in India and developed by the National Environmental Engineering Research Institute (NEERI), Nagpur, India. The technique is available for commercial application and is applicable for all types of mine dumps including the reclamation of wasteland.

Source: Juwarkar, A.A. et al., *Environ. Monit. Assess.,* 157, 471, 2009; Juwarkar, A.A. et al., *Int. J. Min. Reclam. Environ.,* 29, 2015.

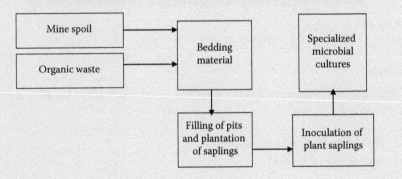

FIGURE 6.7
Process of bio-remediation of mine spoil.

Problem 5

Problem of water pollution due to scree flow in the water channel

Solution

To prevent the flow of scree along with mine water in the water channels, check dams should be provided at regular intervals. Provision of such interrupting structures reduces the velocity of water and prevents the scree flow, but they often get damaged. This problem is further aggravated in the steeper slopes, that is slopes of more than 25°, and could be overcome by the adoption of improved designs of check walls or check dams with filtering arrangements, as shown in Figure 6.8. These check

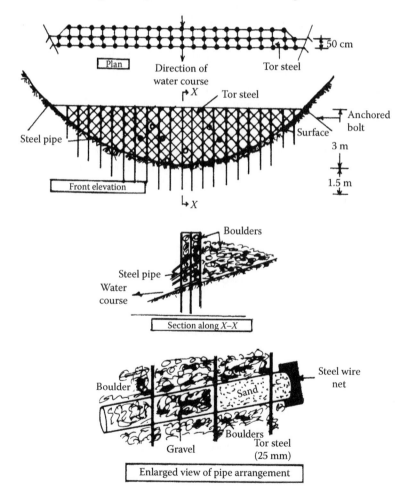

FIGURE 6.8
Design of check dams/check walls with filtering arrangements to prevent scree flow and control water pollution.

dams can also be built easily from the scrap steel available at the mine site and are cheaper than normal wall-type design. Such structures are strong enough to withstand the impact of rolling boulders and will also arrest silt from water.

Problem 6

Lack of environmental-friendly equipment/machinery and accessories

Solution

Though small-scale manual mining does not involve the use of heavy machinery, in cases where drilling and blasting are to be eliminated, the use of rippers as an environmental-friendly means is the best and suitable for rippable rocks and medium-to-low production; for example, limestone is a sedimentary rock that is rippable if it is fractured. For a cluster or group of small mines, the use of rippers is economical and productive.

Blasting operation accessories and equipment: To control the problem of fly rock, ground vibrations and the damage to house and cracks in buildings near the mine areas, the use of a 'sequential blasting machine' (SBM) (Box 6.2) as a environmental-friendly means could be adopted in small mines, and its use can be made for a cluster of small mines. The use of instantaneous detonators is in vogue in small mines, which has detrimental effects on the buildings and structures in the vicinity due to vibrations. It is possible to overcome this problem by the use of either an SBM, delayed detonators or electric detonators for routine blasting. Controlled blasting is yet another solution for alleviating the negative impacts on the environment and also for the problem of fly rock and ground vibrations.

The use of heavy earth moving machinery (HEMM) and equipment on a sharing basis is economical and productive for a group of small mines. It is environmental friendly, too, if used judiciously.

Problem 7

Improper mine drainage causing water pollution in the river channels down below

Solution

Diverted surface water entering into the mine from the adjoining higher altitude areas, together with the drainage water from the mine into the valley, causes siltation and water pollution in channels around the mining areas. Drains could be provided along individual benches, and these drains can ultimately terminate in the main drain along the mine boundary. To prevent soil, waste and debris flowing along the hill slopes, gravel or sand-packed pits, as shown in Figure 6.9, can be constructed either in a number of places or on two extreme ends from where the water is discharged to the hill slope. Drains to be constructed along the hill slope can extend up to the valley

BOX 6.2   WHAT IS A SEQUENTIAL BLASTING MACHINE (SBM)

In conventional blasting procedure, millisecond delay detonators (25 ms delay) are used, which has a fixed delay interval/time. Thus, the flexibility of varying the delay period between holes in a row or holes in a different row is severely restricted. If we can make use of a device that can permit easy variation of the delay interval between two successive holes, then the blasting efficiency can be enhanced considerably and the problem of blasting hazards can be kept under control. This can be done by SBM, which is basically a battery-operated capacitor discharge exploder with the ability to explode 10 different series of detonators at a programmable time interval ranging from 999 ms with the least possible increment of 1 ms.

ACCESSORIES REQUIRED

(with model no. BM-175-10-PT of Research Energy of Ohio, USA)

- *Extension cable*: A multicore cable for connecting 10 circuits to sequential blasting machine and terminal board.
- *Terminal board*: It has 10 pairs of terminals for 10 circuits.
- *A0-5-load plug*: It is used to test sequential blasting machine. The load plug contains 10 resistors, each equal to the maximum load resistance of blasting machine.
- *CR-50 cable reels*: An aluminium reel used to carry and transport explodable cable.
- *ET-175-10 Energy Tester*: Provides nominal test values for energy output (measured in and percentage of rated energy for each of the 10 circuits of sequential blasting machine.

ADVANTAGES OF SQM

- Permits larger blast of 20,000–25,000 tons size also.
- Controls ground vibration and noise.
- Improves fragmentation through precise blasting, as there is flexibility in changing the delay timing between holes in a row and different rows.
- Separate heaps of different grades are possible with proper sequences of delays.
- The machine is safe during use as it is provided with a safety interlock.

- By trial and error, with varying delay time, the technique can be perfected to local geological conditions specific to the fragile Himalayan region.

Though other blasting methods (using delay detonators) are available for large production blasts, SQM is more suited for small- and medium-scale blasting in a environmentally sensitive area and in a controlled manner, where optimum results can be achieved through trial and error.

Source: After Singh, S.K., Use of sequential blasting machine for deep hole blasting at Gagal works, *Souvenir Volume of Mines Environment and Mineral Conservation Week,* 1995–96, Organised by IBM, Dehradun and Mine Lessee of Dehradun Region, Dehradun, India, 1996, pp. 67–68.

depending on the requirement. Other ameliorative measures to prevent water pollution during operational phase of the mine include

1. Construction of a series of toe walls made of rubble
2. Plantation of shrubs, grass and saplings on the slopes to check soil erosion

Problem 8

Improper biological reclamation of mined-out area and dump sites in small mines

Solution

This problem is due to the lower survival rate of plants. Even local species, which are preferred for reclamation work in mines, are unable to grow. It should be noted that 'biological reclamation' of derelict land should first increase the organic value and water holding capacity of the land for plant survival, which in turn supports the growth of vegetation because biological reclamation involves more than mere planting of trees. It starts with grass, followed by shrubs and trees later, as this type of succession increases the overall fertility of the soil and ground in a permanent manner. The use of fertilizer and chemicals to improve the fertility is a temporary measure. The following principles need to be applied with the integrated approach:

1. In all plantation works, preference should be given to local species than exotic ones and mixed cultures in preference to monocultures (Table 6.4).

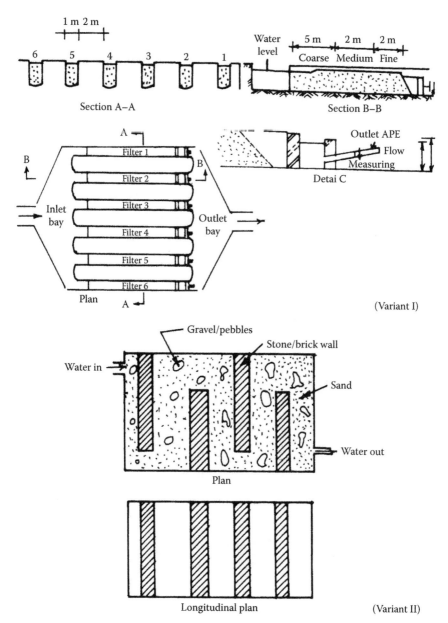

FIGURE 6.9
Gravel/sand-packed pits.

2. Some selected species which can grow quickly, stabilize effectively and improve the soil quality with time, besides being economically useful, can be planted for reclaiming the wasteland generated by mining.

3. In areas which are directly visible, some cosmetic plantation can be done for early results, thereby hiding the scars of the damaged land.

4. The use of microorganisms and other organisms living in the soil should be adopted for enhancing the fertility of the land in a natural way.

MINING PROBLEMS

Problem 1

Ugly scars created as a result of mining are an eyesore and give effervescence to environmental hullabaloo.

Solution

Design the mine faces/benches in such a manner as to form a 'bowl-shaped mine' after its full development. Though this may not be feasible for all deposit forms, wherever feasible, the planning should be done accordingly. The visibility of ugly scars due to mining can be reduced with this approach.

Problem 2

Orientations of the working faces are not proper.

Solution

1. The orientation of the working face should be kept in the direction of strike of the deposit and opened along the contours (not along the spurs) to create larger space availability.

TABLE 6.4

Some Agro-Forestry Systems in the Himalayas

(A) Tree Crops	
Tree Crop	Intercrop
Kashmir	
Almond/apple (rows)	Saffron, vegetables
Walnut (on drains/bunds)	Maize, beans
Robinia pseudoacacia, Celtis, Ailanthus, Morus, Salix, poplar	Paddy, wheat
Leucaena, mulberry, poplar	Paddy, wheat
Acacia nilotica, eucalyptus, citrus	Pulses, maize, fallow

(Continued)

TABLE 6.4 (*Continued*)

Some Agro-Forestry Systems in the Himalayas

(A) Tree Crops	
Tree Crop	**Intercrop**
Himachal Pradesh	
Plum/Apple	Vegetable, wheat, maize
Pinus roxburghii	Elensine
Kydia, Ehretia, Morus, Grewia	Paddy, wheat, gram
Uttarakhand	
Kydia, Grewia, Morus, Terminalia, Celtis, Ficus, Ougenia, Prunus, Pyrus	Paddy, wheat
Tarai	
Poplus deltoids, eucalyptus	Sugarcane, paddy, potato
Leucaena, Sesbania, poplar	Brassica, cynleo
Natural forest	Maize–wheat–paddy–wheat, coffee, cardamom, rubber, dalchini, smilax, costus, etc.; *Solanum khasianum, Dioscorea* sp., *Gynocardia odorata, Strychnos nux-vomica*
Arunachal Pradesh/NE Himalayas	
Dipterocarpus macrocarpus, Terminalia microcarpa, Shorea robusta	Intercrop of cocoa, coffee
Grevillea robusta, Alnus nepalensis	Tree, cardamom, turmeric, ginger
Indigofera tasmania, Albizia lebbeck, A. procera, A. moluccana, A. sumatrana	Tea
Terminalia microcarpa, Bischofia javanica, Alnus nepalensis	Cardamom
Dipterocarpus macrocarpus	*Piper nigrum*
Dipterocarpus macrocarpus	Ginger, *Coptis teeta*
Assam/NE Himalayas	
Areca	Black pepper, banana/papaya, as filler crop
Areca and citrus	Pineapple, betel vine, cardamom,
Grevillea robusts	Paddy
Bamboo, *Parkia roxburghii, Mangifera indica, Syzygium cumini, Emblica officinalis, Morus alba*	Vegetables (chilies, beans)

(B) Fruits		
Name of Fruit	**Botanical Name**	**Improved Varieties**
Apple	*Malus sylvestris*	Red June, Summer Golden Pippin, Fanny, Jonathan, Red Delicious, Royal Delicious, Golden Delicious, Granny Smith, Maharaji Rich-a-Red
Pear	*Pyrus communis*	Leconte, Kashmiri, China, Bartleet, Smith, Gola, Victoria, Fertility

(Continued)

TABLE 6.4 (*Continued*)

Some Agro-Forestry Systems in the Himalayas

(B) Fruits		
Name of Fruit	**Botanical Name**	**Improved Varieties**
Peach	*Prunus persica*	Flordasun, Elberta, Alton, JH Hale, Crawford's Early Koora, Nectarine Sun Red
Plum	*Prunus domestica, P. saliciana*	Santa Rosa, Satsuma, Excellsior, Doris, Laddakh, Kelcy's Japan, Grand Duke, Beauty, Alpha, Red Ace, Wickson
Apricot	*Prunus amygdalus*	Negette, New Castle, Saffaida
Almond	*Prunus amygdalus*	Negette, Kagzi Budded, Non Pareil Drake, California Paper Shelled, Thinshelled
Walnut	*Juglans regia*	Govind, Kagzi, Tuttle 31, Ideal
Sweet orange	*Citrus sinensis*	Mosambi, Blood Red Malta, Jaffa, Washington Navel, Pineapple, Valentia Late
Lemon	*Citrus limon*	Eureka, Lisbon, Pant-1, Assam
Mandarin	*Citrus reticulata*	Coorg, Sikkim, Khasi, Kinnow
Grapefruit	*Citrus paradisi*	Marsh Seedless, Duncan, Ruby, Marsh Pink
Pummelo	*Citrus maxima*	White Fleshed, Red Fleshed
Guava	*Psidium guajava*	Allahabad Safeda, L-49, Red Fleshed, Chittidar, Karela, Pear Shaped, L-42, Seedless, Apple Colour
Anola	*Emblica officinalis*	Banarsi
Jackfruit	*Artocarpus heterophyllus*	Local varieties
Jamun	*Syzygium cumini*	Local
Fig	*Ficus carica*	Local
Bael	*Aegle marmelos*	Local
Pineapple	*Ananas comosus*	Kew, Queen, Giant Kew, Mauritius
Passion fruit	*Passiflora edulis* var. *edulis*	Purple fruited
Pecan	*Carya illinoensis*	Hachdiya
Persimmon	*Diospyros kaki*	

Source: After Gupta, R.K. and Arora, Y.K., Technology for the rejuvenation of degraded lands in the Himalaya, in: Singh, J.S., ed., *Environmental Regeneration in Himalaya: Concepts and Strategies*, Central Himalayan Environment Association, Nainital, India, 1985, pp. 169–186.

2. Hill areas face high wind speed most of the time, and working in such high wind speed is difficult. Therefore, at the time of planning, the orientation of the face should be selected in such a manner that the face lies along strike and is also protected from the wind.

3. While opening mine faces along hill slopes, those sides of the slope that face the sun during the hottest period of the day should be chosen. It makes working conditions in cold weather easier and also improves the growth of vegetation naturally, thereby making reclamation easy.

Problem 3

There is inadequate bench height and width in the small mines according to the statutory requirements of mining laws.

Solution

The permissible height of benches in small open mines is 1–1.5 m under the existing Indian laws. To maintain better productivity, heights up to 4–6 m can be easily planned, but the statutory requirements of mining laws do not permit more. Thus, this point can be made flexible according to the site conditions, thereby bringing improvement in the mining conditions. According to the bench height, the width can be planned.

Problem 4

Rolling of rock boulders down the slopes and safety of the persons.

Solution

As a general practice, in hill mining, small quantities of rock/mineral are left at the hill edges, called the 'toe' (especially in a hilltop open-strip mine). If the extraction is done by the hilltop chopping of the deposit, then no toe should be left at the edges of the hills, as it can cause rolling of boulders down the slopes, endangering the safety of the person working below.

Problem 5

Inadequate mine planning on a long-term and short-term basis exists.

Solution

Planning and development are the two important and interwoven stages of mine planning. Whether a mine is small or big, long-term as well as short-term developmental planning is essential for scientific mining.

Long-term planning is required for opening the deposit, advance planning of bench faces for production, reclamation of mined-out area, de-commissioning and so on. It should be carried out in a phased manner right from the inception, as this will enable the design of the mine in a systematic manner. The planning for transportation, waste management and so on should be done from a longer perspective, and for that also long-term plans are required. However, day-to-day developmental work and production target achievement require short-term planning. It is important that short-term plans are synchronized with the long-term objectives; otherwise, unscientific and haphazard development of the mine will result.

Private and small entrepreneurs engaged in mineral extraction do not pay attention to this vital yet important practical aspect, as they want to follow shortcuts. They should avoid shortcuts and plan for both short term and long term.

Problem 6

Unstable overall pit slopes in small mines cause destabilization of the slope.

Solution

Slope stabilization of mine faces and other slopes is an inherited problem in hill mining, and it is further aggravated when the rocks are geologically weak. It is therefore necessary that an 'optimum slope angle' for various rock types must be determined. The overall optimum slope angle (σ) can be determined at the planning stage and should be utilized for the total life of the mine. The suggested slope angles are given in Table 6.5.

The slope angle (σ) is a function of the structural, hydrological and strength parameters, such as rock types, dip of the formation, thickness of bed rocks, cohesion, jointing pattern of the rock, water and permeability of rocks.

Problem 7

Approach road and haul road designs are unsatisfactory and marks with basic deficiencies.

Solution

1. A 1 m high projected parapet wall along the road edges is needed for hilly approach roads and haul roads.
2. At the curves, super elevations are necessary.
3. The surface of the road should slope inwards and be provided with suitable drains for surface water run-off (Figure 6.10).
4. Road layout design should be done taking into consideration the hanging wall and foot wall (HW and FW) side of the mineral body, as it shortens the overall length of roads.

TABLE 6.5

Stable Overall Pit Slope for Various Rock Types

S. No.	Type of Rock	Stable Pit Slope (°)
1.	Soft clay	25–35
2.	Compact clay	30–40
3.	Hard shale, sandstone, limestone	40–45
4.	Hard sandstone, hard limestone and dolomite	40–45
5.	Weathered igneous rocks	40–50
6.	Very hard sandstone, limestone and dolomite	50–60
7.	Metamorphic and igneous rocks	50–60
8.	Very hard metamorphic and igneous rocks particularly hard quartzites	60–70

Source: Soni, A.K., An appraisal of environmental problems of mining in Himalaya with possible remedies, *International Conference on Impact of Mining on the Environment: Problems and Solutions*, VRCE, Nagpur, India, 1994b, pp. 253–266.

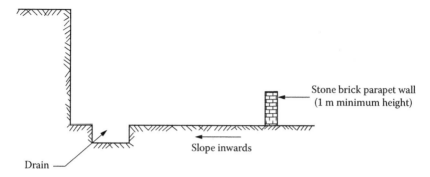

FIGURE 6.10
Cross-section of the road and parapet wall.

TABLE 6.6

Angle of Slope and Percent Grades for Various Popular Truck Models of Tata Make

			Maximum Climbing Ability	
S. No.	Model	Make	In % Grade	Angle of Slope (Approximate)
1.	SE 1510	TATA	13.7	7°41′
2.	SE 1510A	TATA	12.7	7°8′
3.	SA 1210 (4×2)	TATA	18.0	10°12′
4.	SA 1210 (4×4)	TATA	27.2	15°16′

5. For approaching the deposit, hill roads with a minimum road width of 6–8 MT are necessary, which may be constructed with a locally available material. But, in the hill, it is not possible to make the roads of the required width at all places. Wherever such constraints exist, the minimum width for single-lane traffic should be maintained.

6. The road gradient permitted by mining laws ranges from 1:10 to 1:16 in case of open-cast mines. The real condition in the hilly topography is different because hill roads are generally steeper than the normal roads of flat topography. If the gradient is to be maintained according to the law, the length of the roads becomes too much, thus rendering the transportation costlier. Therefore, it is necessary for mine roads in the hills to have a more acute gradient. Table 6.6 gives the manufacturer's specifications in terms of equivalent angle of slope and maximum climbing ability as the percentage grade and angle of slopes for one popular truck brand (TATA) that is commonly used for limestone transportation in the mines. From this table, it can be seen that with the available trucks, still steeper gradients can be planned.

Hence, the statutory provisions for approach roads, haul roads etc., should be designed accordingly and especially considering the arduous working condition of the mines of the Himalayan area.

Problem 8

The problem of disposal of waste or overburden and its unscientific handling has to be resolved.

Solution

In Himalayan mines, the overburden disposal (waste disposal) is a severe (probably the most critical) problem due to constrained space. The problem becomes accentuated because of the restricted space in the hilly topography, vegetation, steepness of hill slopes and heavy monsoon run-off. In hills, it is common practice to use the deep valley and hill slopes as the dumping grounds. Sometimes, the waste is heaped along the road side and waste is allowed to roll down the slopes. The method of waste disposal along the slope, which is very common, is not the correct method, and it is unscientific too. It should not be adopted for waste disposal. As a general practice, in most of the mines of the hilly area, the scree flow on slopes is a common matter of concern from the environmental point of view. This can be easily prevented by creating an artificial garland-like wall of 1–1.5 m width along the contour of the hill using overburden boulders or rejects from the mine (Figure 6.11). Alternatively, to contain the scree within the mining area, a 'trench on each

FIGURE 6.11
Garland-like wall of overburden boulders/rejects (1–1.5 m wide) in a mine to prevent scree flow down the slope and along the hill contours.

or subsequent benches' of about 1–1.5 m deep and about 0.5–1 m wide could be constructed. The trench would run along the length of the bench in the direction of strike and should be considered as the standard preproduction development work.

The low-cost drum debris retaining wall for slope stabilization is yet another alternative which can be used (Case 4.2.8) in place of a 'stone parapet wall'.

Scientific methods of overburden/waste handling: The study of the dumping sites of various mines has indicated that dumping leads to environmental hazards such as creation of ugly scars, obstruction in vegetation growth on slope and chocking of water channels. The environmental dimension of this particular problem is growing day by day. By adopting BPM, this can be handled easily. Two practical methods of overburden/waste disposal, as described later, are recommended for common mining practices. It can be easily articulated with the integrated approach.

Method I: Waste disposal by backfilling is the most appropriate method that could be adopted with the integrated approach (Figure 6.12). As the name of the method indicates, dumping of reject or waste is to be done in the worked-out area of the pit, thereby confining the waste to the mining premises only. The method of backfilling is speedy and cost-effective, as the transportation distance is small. Transportation can be accomplished either by conventional truck/dumper system or by other systems of transportation such as an aerial ropeway or conveyor. The essential prerequisite for this method is that the provision of waste backfilling should be made from the initial planning stage and continue according to the designed layout plan of mine.

Method II: Another method suitable for waste disposal is the method of 'fill construction in lifts', which can be chosen for reject handling if the topographical features allow. This situation arises wherever limited space is available for the disposal of waste (Figure 6.13). The hill slope can be converted in the form of steps of about 1.5–2 m height by erecting stone walls right from the valley level. An approach road is required to be developed up to the valley. Erection of stone walls can be done by using wire nets of 25 mm × 25 mm size.

For hilltop mining (both open-pit mine and open-strip mine), a typical mine waste disposal system is illustrated in Figure 6.14.

Problem 9

Production overshadows SHE (safety, health and environment), which is neglected especially in small mines by private entrepreneurs.

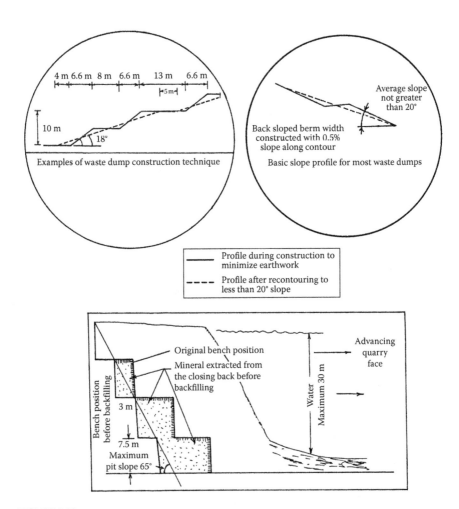

FIGURE 6.12
Waste disposal by backfilling. (After IBM, Environment aspects of mining area, Bulletin No. 27, Training, Mining Research and Publication Division, Indian Bureau of Mines, Nagpur, Publication of Ministry of Mines, Indian Bureau of Mines (Government of India), Nagpur, India, 1994.)

Solution

It is beyond doubt that safety is a major concern for the entire mining industry and required by all mining professionals. We all want to live in a clean environment and go back from work safely and without inviting any health troubles. Willful ignorance of SHE is true especially with private entrepreneurs. If safety standards are brought down, industrial operation may become unsafe and unproductive. Unsafe practices lead to loss of time, money and employee's morale as well (Soni and Kiran, 2012). Thus, safety and productivity go hand in hand.

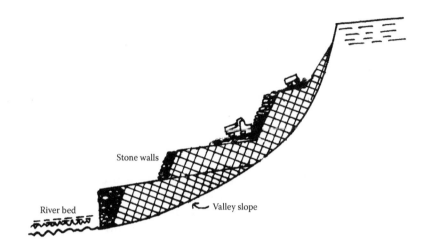

FIGURE 6.13
Method of fill construction in lifts.

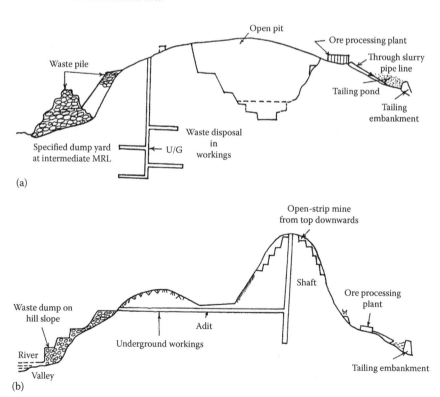

FIGURE 6.14
Typical mine waste disposal system. (a) An open-pit mine, (b) An open-cast mine. (After Soni, A.K., Integrated strategy for development and exploitation of mineral resources of ecologically fragile area, PhD thesis (unpublished), Indian School of Mines (ISM), Dhanbad, India, 1997, p. 238.)

Safety in the Himalayan mines can be enhanced or dealt effectively with knowledge of safety management and safety engineering, which are the modern and newly emerged tools. There are three basic guiding principles in this context:

1. Known support of the management (at all level of workers and management)
2. Real commitment at every level (with action and not verbal)
3. Individual empowerment (to correct unsafe conditions and practices)

Accidents are preventable, and safety can be ensured by organized approach and involvement of all in the hierarchy. For effective results, the study of all working areas, all operating methods and practices, to detect, eliminate and control physical hazards that contribute to accidents can help. Education, instruction, training and discipline, which identify unsafe acts or accidents, can determine the contributing circumstances. Thorough investigation of incidences can minimize human factors to achieve zero accident.

SHE should be given the highest priority for fragile areas, and importance should be given as per the conditions encountered.

BPM, as previously described, together with the new concepts and eco-friendly mining solutions, which require due recognition in the twenty-first century, is extremely useful for the mineral exploitation in the Himalayas. Various solutions with respect to mines of the Himalayas have been tabulated in a concise form and are given in the following. This is based on the observations described and experience gained over the years. In this way, environment and mining can move together for sustainable development of the country in general and the Himalayan region in particular.

Observations/Problems	Eco-Friendly Solutions
1. No detailed exploration data available	• Exploration by modern methods and sophisticated instruments • Computerized advanced data analysis for accuracy • Application of cluster mining concept for SSM
2. Dominance of unscientific mining practices in SSM sector particularly	• Planned scientific mining from hilltop to bottom for domal type of deposits • Systematic exploitation in the form of benches by maintaining overall bench slope according to the rock type

(Continued)

Observations/Problems	Eco-Friendly Solutions
	• Planned mine development including appropriate face height and width, regular and even working faces
	• Selective mining in case of need
	• Avoiding mining along the roadside, etc.
	• Apply cluster mining concept for SSM
3. Conventional ROM transportation in hill is more costly and eco-unfriendly	• Eco-friendly alternatives for ROM transportation from hilltop to bottom or from one hill to another by the following methods:
	• SFRC chute
	• Vertical silo with conveyor arrangements
	• Shaft–adit combination (SAC method)
	• Ore slides in rock
	• Overland belt conveyor system (OLBC)
	• Gravity-operated skips on guided rail system (GOSS)
	• Gravity for transportation down the hills
	Note: If cost is not a major hindering factor, make use of innovative and new techniques, for example snake conveyors or truck lift arrangement. In the long run, its indirect returns are immense. These are economical and safer for sensitive environments also.
4. Lack of minimum machinery required for systematic mining	• Rippers
	• Hydraulic breakers
	• Hydroseeder for reclamation work on steep slopes
	• Sequential blasting machine, use of delay/electric detonators, etc.
5. Safety in mines and mining areas	• Adoption of a zero-accident approach
6. Severe and visible mining-induced land degradation problems, including soil erosion problem, of hill areas	• Application of IBA (integrated biotechnological approach) for reclamation of mining site and spoil dumps, etc., in a permanent way
	• GIS application and remote sensing methods for the analysis of environmental data and environmental management of mining operations including land degradation studies
	• Soil conservation techniques using natural and artificial methods, for example coir mats and geotextiles
7. Water pollution and scree deposition problem	• Cost-effective and innovative design of check dams using locally available material to prevent scree flow and control water pollution
	• Concept of gravel/sand packet pits

(Continued)

Observations/Problems	Eco-Friendly Solutions
8. Air pollution and fugitive dust emission due to mining	• Eco-friendly approaches for dust suppression at the source of generation and adequate safety equipment for air pollution hazards
9. Lack of synthesis of social and industrial need	• Integrated approach with BPM • Participatory management approach
10. Legal and financial problem for small and medium entrepreneurs	• Recognition of SSM as private industry and financing facilities with financial institutions • Mining according to a recognized mining plan and environmental management plan (EMP) • Framing of guidelines and their upgradation from time to time as per requirement
11. Competent skilled/unskilled manpower	• Adequate human resource development and training

6.4 People's Participation in Environmental Management

With reference to mining industry in India, labour and the environment are the two important constituents which draw the attention of the public at large. These two most vulnerable components of industrial activity generally culminate in industrial disputes/unrest. Hence, community consultation and its active participation are extremely important for the implementation of planned environmental measures. It must be inculcated in the minds of people that environment or enviro-oriented development of mining areas is not a task of the company or organization alone that is engaged in mining but it is also a people's cause and needs cumulative action(s). BMP suggests the need for people's participation in management at various levels for both mining and the environment. Though people's participation in mining management is not new, in respect of environmental management in the Himalayas, it is as essential as the other factors. Community consultation and its participation in environmental management basically explain to the public

- How the company is beneficial for local population in particular and for the nation as a whole
- How and in what manner an individual can gain from the company's positive practices
- What the company's plans are for environmental protection and mineral conservation
- What the company's plans are in future for the social development of the area

Some other important and pertinent questions about the discussed topic are the following: who should be involved, where to begin and how often such participation should take place. However, there are no fixed rules about such participation of community, and these are entirely based on the prevailing need and human response. The objective should be to create a win–win situation for both.

6.5 Role of Organization

Preparation of an EMP, detailing how the mining company intends to manage and reduce impacts, is an important task, but its actual implementation in the field (or in the mine) is a more important component, as emphasized by the BPM. It is a relevant adage that a 'good start means half done', and here, the role of the organization is significant. Early planning identifies most of the mining and environmental issues that need to be managed. In addition to the implementation of management measures, BMP suggests placing emphasis on environmental auditing and human resource management by training.

The role of the organization is very important in providing organization leadership, which drives its culture of work and help to develop, support and improve environmental performance. Some broader and noteworthy points of interest in this context are (1) environmental-friendly mine planning, (2) environmental management of mining operations and (3) clean technology promotion, for example environmental-friendly transportation. With these guiding principles, environment-oriented development in mining areas of hills can be successfully achieved.

7

Environment-Oriented Development

By interrelating the concerns of safe mining, mine productivity and other region-specific parameters, as described in the earlier chapters, it emerges that 'environment' is the basic link in the development of the ecologically sensitive Himalayan region. An integrated strategy and a multidisciplinary approach for planning in the Himalayas can minimize the lacunae already identified. In such sensitive areas, mining and the environment should be considered together; specifically, the strategy should be an *environment-oriented development* (Moddie, 1980).

The Himalayan mountain environment and its management has emerged as one of the most challenging tasks for both individuals and organizations in the twenty-first century. A continuous improvement approach based on the plan–do–check–act strategy can be adopted to achieve better results. This basic principle can lay the foundation for the success of the overall mining and environmental management operation. However, the basic needs of food, water, fuel, energy, fodder and shelter for hill people are imperative and have to be met on priority for satisfying human requirements.

Advanced planning and its enforcement, monitoring of mining and environmental problems and its supervision, together with scientific research and its application, are some recognized prerequisites for the development of mining activity in the Himalayas. Mining ventures in the Himalayas may lead to confrontation between developers and environmentalists. Teamwork, through the cooperation of industry, government and other stakeholders, can protect environment and serve as a model for peaceful coexistence of the mining industry and the environment. It has been experienced and realized that this approach can make the overall development successful.

7.1 Development Imperatives

The basic objectives of environment-oriented development of the mineral-bearing areas in the Himalayas are as follows:

- Protection of environment and ecology
- Protection of biological diversity and maintenance of essential ecological processes and life support systems

- Enhancement of the overall standard of living by following a path of economic development that safeguards the welfare of future generations

Here, it may be reiterated that *sustainable development, best mining practice* (BMP) and *integrated approach,* as described in previous chapters, for implementation is the green development agenda for the Himalayas. The inherent mistake of treating 'hill development' as an extension of plain development should not be made. Optimum performance, within the limits of an achievable and acceptable environmental protection level as applicable to the fragile/sensitive areas, can be attained easily through these concepts.

A comparative review of the Himalayas and other mountainous areas of the world is, though theoretically desirable but practically unfeasible for implementation, as the defining parameters sharply differ from one region to another; for example, no uniform parameter exists for 'fragility' all over the world. However, the reader should know and understand the environmental consequences thoroughly in the context of mining or the mechanism of mineral extraction. Such knowledge is helpful and essential for the effective management and formulation of future strategies. The fragile hill slopes of the Himalayas should be understood, compared and evaluated by identifying the technology imperatives, operation of mines (world standards vs. local conditions) and environment and ecology protection from the view point of potential scope of hill development. Other parameters such as weather/climate, tectonic activity, fragility/sensitivity and geological setting, which matter significantly, should be compared on a case-to-case basis as they are site specific. At every stage of mining, a comprehensive environmental monitoring system together with an integrated and interdisciplinary approach is thus suggested for environment-oriented development.

7.2 Affected Environment

The principal goal of mining is to carry out mineral winning and mineral benefaction activities in an environmentally sound and safe manner. Possible impacts of mining operations on the environment and community may or may not be significant. Hence, an essential starting point to describe the affected environment is to have the base line data (pre and post mining) of local environmental conditions, knowledge of accepted best environmental management practices and the possible consequences to the environment resulting from the mining activities. The evaluation of the possible effects on the environment, commonly referred to as 'environmental impact assessment' (EIA) or preparation of an *environmental impact statement* (EIS), should be done under two major heads:

1. Environmental performance assessment by field monitoring
2. Scientific analysis of the monitored data

Such descriptive and analytical analysis contains information about the environmental setting of the study area, which forms the basis to know whether the mining process is causing any effect on physical, biological and social systems. It also informs about the status of the environment before and after mining. Large-scale mining projects require detail EIA/environmental management plan (EMP) studies, and small-scale mining projects usually require a review of the environmental factors. Though the complete procedure for the description of the affected environment (Canter, 1996) in a few pages is difficult, briefly this can be accomplished in the following manner:

7.2.1 Environmental Performance Assessment

Environmental performance assessment is based on the environmental monitoring carried out in the field and is a measure of the success of the strategies implemented. The diversity of climate, ecosystem, land use and topography greatly influences the design of an environmental monitoring program and thereby performance assessment. In environmentally sensitive regions, this monitoring program should be comprehensive enough to represent the entire area. If done so, it will also enable the review and improvement of the management plans or measures. The environmental performance needs to be monitored periodically against the following objectives set out in the environment management plan (EMP):

- To detect short-term and long-term trends
- To find out the causes of environmental degradation
- To improve environmental practices and procedures
- To demonstrate to the community/government that the operation complies with environmental quality/standards
- To assist in regional or micro-level planning
- To know the premining and postmining status

Though environment performance assessment is a broad and very specialized area, for BMP it encompasses aspects related to air, water, land, biology, noise, vibration and ecosystem protection. Standard operating procedures (SOPs) for field and laboratory should be used.

7.2.2 Air Quality

The monitoring and assessment of air quality are a major component of the environment performance assessment. Ambient air monitoring and air monitoring at the work sites are two different areas in respect of air environment assessment. Dust and fumes (blasting) are two by-products that deteriorate the overall air environment in any mineral extraction projects. Air pollution and its monitoring at the project site, where the actual operation goes on, need to address the following requirements:

- Location of the monitoring site
- Frequency of observations
- Equipment to be used for monitoring and data collection
- Appropriate quality control procedures to ensure the reliability of the results

Both laboratory and field data should be analyzed to obtain the most representative results of air monitoring. High-volume air samplers and respirable dust samplers are the instruments required in the field for sampling particulate matter as well as gaseous pollutants such as SO_2 and NOx, which are the prominent indicators of air quality. If needed, other sampling analyses (e.g. radiometric and gravimetric) can be carried out depending on the project requirements.

7.2.3 Water Quality

In order to assess and monitor water quality, it is necessary to know the characteristics of the natural water available in the area and what might be introduced into it as a result of the mining operations. The likely impact of mining on the water quality in mountain areas may be due to one of the following:

- Contamination of water channels in the vicinity of mine (pollution due to effluent or run-off from the mine site)
- Contamination of water springs and groundwater

The chances of groundwater contamination are very small in such areas because of the very nature of the mining process and for obvious topographical reasons. Moreover, groundwater table lies at a considerable depth and is highly fluctuating according to the hill profile. It is observed in some limestone mining areas of Sirmour district in Himachal Pradesh that natural springs occurring in the mining areas have dried up. This may be due to the change in the rock fracture pattern of the area or to the change in the hydrological regime of the area. Hence, it is clear that drying of springs or changes in water channel courses may occur as a result of mining.

To detect the pollutants present in mine water, more than 35 standard parameters, as laid down under the Water Prevention and Control of Pollution Act, must be evaluated (Table 5.A.3). Water quality is assessed by analyzing samples in the laboratory, which enables obtaining the key indicators of water pollution. Special attention must be paid to the sampling method and to the preservation and handling of water samples prior to analysis. It should be done according to the SOP, as indicated earlier. Water pollution parameters such as pH, BOD, COD, turbidity, DO and coliform level must be determined with high precision before discharging the mine water into mountain water channels, that is rivers, streams, in the valleys for the protection of aquatic life forms.

Since water is a major transport medium for contaminants, a judicious water monitoring programme needs to address the following requirements for best practice mining in ecologically sensitive areas:

- Water sampling points and their location should be such so that they truly represent the water quality of the area.
- Sampling frequency should be in tune with hydrological variability and seasonal requirements.
- Standard sample collection, preservation techniques (before analysis) and analytical methods of chemical analysis should be adopted.
- The test results should be evaluated, and monitoring progress and/ or practices should be reviewed.
- Appropriate and superior quality control procedures/management measures should be implemented.

In addition, siltation of water channels, which becomes problematic in hill areas and may cause pollution, should be prevented at the source itself.

7.2.4 Land Degradation Assessment

Land degradation due to mining is manifested by various physical impacts (Saxena and Banik, 1996; Saxena et al., 2002). Land monitoring for its evaluation essentially relates to the land management components of an EMP and is the most important component in mountainous terrains because of the fact that degraded land will look like scars, which is an eyesore (Saxena, 1995). Best practices with regard to land include the quantification of land degradation, that is change detection analysis over a time period (Soni and Loveson, 2003). A land monitoring programme needs to address the following:

- Identification of areas to be monitored for each of the key issues
- Specification of the appropriate method and frequency of monitoring for each aspect
- Evaluation and review of the results/observations and adjustment in the monitoring progress and/or practices

Geographic information system and remote sensing techniques can be used for land degradation assessment, as the mountainous terrain is not easily accessible for field data collection. Environmental information (from primary and secondary data sources) can be derived from field survey and soil map, watershed map, slope map, land use map, land categorization map, vegetation map, etc., can be prepared in the electronic form. Land maps prepared in this manner can assist in the quantification of land degradation over a period, such as topsoil/subsoil management, erosion control and landslips/ slides management, and for the protection of aesthetic features of hills.

For hilly areas, land management based on a land capability classification approach (Soni, 2002) is an important tool, especially for sustainable land development in mineral-rich areas.

7.2.4.1 Restoration of Derelict Land

Mined land is basically derelict land, which has lost its organic content and water holding capacity. Its restoration to original or near-original shape can be done by appropriate treatment and aftercare. Two methods, namely natural restoration and restoration by plantation (the IBA approach), are the most adequate and suitable methods for making the derelict land 'green', but social compulsion and industrial needs of the mountain areas also provide an 'additional alternative use' of derelict mining land. This includes restoration for 'miscellaneous purposes' such as the development of recreational centres, tourist resorts, playgrounds, cattle grazing grounds, animal orphanages and animal sheds. Restoration of derelict land in mining areas of mountains requires the following measures in a phased manner (three phases).

Phase I
- Collection of complete information or microclimatic conditions of the area which include rainfall, temperature, humidity and snowfall
- Collection of the contour map of the area
- Collection of physico-chemical characteristics of the soil, such as pH; organic/humus content; K, N and P content; soil depth and texture
- Collection of base line data to check soil erosion and water conservation
- Collection of vegetation records and information about species diversity, community structure and site-specific soil–plant–animal interrelationship of the area prior to the mining activities

Phase II
- Immediate measures to check soil and water conservation (if required).
- Improvement of land quality by additives, fertilizers or manure.
- Choice of appropriate local species. Selection should follow the basic ecological approach. Give preference to local species and plant

species that are capable of growing on barren, rocky areas deficient in moisture, nutrients and organic matter.

- Adopt other land management measures taken locally. The practices adopted should be such that they help improve the land quality in the long term.

Phase III

- Ground to be levelled, compacted and terraced.
- Wherever permissible, tillage operation can be used.
- Mechanical means of removing stone may be adopted if the soil to stone ratio is high.
- Nitrogenous fertilizers are needed for derelict land at the initial stage of land quality improvement. Legumes grown on mine soils are likely to increase the amount of soil nitrogen, as mine soils are low in organic matter and nitrogen. To acquire organic matter, tree leaves may be allowed to decompose and earthworms can be added to accelerate the process of decomposition.
- Vegetation grown on mined land utilizes the water and nutrients of mine soil for their survival. Plants must have about 20 different nutrients to grow and develop properly. Not all plants use all elements but six elements, namely nitrogen, potassium, sulphur, calcium and magnesium, are necessary. Therefore, selection of suitable fertilizers containing the essential elements must be done. If there is a deficiency of one nutrient needed for plant growth, the mere addition of some other nutrient will not increase or support plant growth. Therefore, only the deficient nutrient is to be added.
- Land management according to land categories, that is class I to class VIII, should be applied (Appendix 7.A) and derelict land should be revegetated to its near-original pristine state.

Restored derelict land area should be protected from grazing, and proper care should be taken for protection from fire and other hazards. The improvement of degraded/denuded land may take many years.

7.2.5 Noise, Vibration, Air Overpressure and Fly Rock

Blasting is an important component of the mining process. The environmental hazards associated with blasting include noise, vibration, air overpressure and fly rock, which are the undesirable side effects. The topography of mountains is a responsive or an automatic multiplier of noise, vibration and fly rocks. When blasting is carried out, it is accompanied by a loud noise. It is an atmospheric pressure wave consisting of high-frequency sound of short duration, which is audible. In environmentally sensitive places such as hill areas, normal blasting operation should be seen from an alert view. In most hill places, blasting causes psychological fear rather than any permanent damage. The detrimental effects of hill blasting are cracks in the window panes of buildings and fly rocks in the valley.

Noise, vibration, air overpressure and fly rocks are well addressed in any mining project report and environment management plan. A number of measures to prevent these hazards have been suggested (CSIR-CIMFR, 2007), but for best practice mining some cost-effective measures should be put into practice, which include the following:

- Proper blast design which includes the placement of explosive, blast pattern shape, point of initiation, subdrilling, burden, spacing and stemming column length
- Adequate charge per delay and proper loading of explosives
- The use of delay detonators and correct selection of delay intervals or alternative use of a 'sequential blasting machine'
- Direction of initiation
- Low-density explosives in loose and fractured rock mass
- Selection of appropriate time for blasting
- Muffle blasting to control fly rocks
- Angular holes in conformity with the slope of the bench
- Controlled blasting techniques (if necessary)

7.2.6 Biological Environment Assessment

The checklist required for the assessment of the biological environment is a long one, which includes the flora and fauna. All vegetation types and wildlife categories, including endangered species, fall under this environment assessment category. Their species density, population, sex ratio, pattern of multiplication/breeding, growth rate, mortality rate, migration pattern, disease pattern, etc., which are affected by industrial activity and human intervention, need to be understood for safeguarding the biological environment. These are to be first identified during the EIA process, and then corrective measures are to be planned and implemented for environment protection and preservation so that adverse impacts are minimized. Because of the complexities and dynamics of biological systems, studies need to be carried out over more than one season in a year. It will define the key indicators that will be used in the assessment process. A biological environment assessment program must be run in conjunction with the ecosystem preservation plan, as they are interwoven (EPA, 1995). A biological monitoring program needs to address the following requirements:

- Define the community and species dynamics.
- Select appropriate indicators for direct toxicity or bioaccumulation.

- Consider variation due to space and time.
- Evaluate the direct impact on biological communities.
- Use widely accepted and standardized methods wherever possible.
- Collect adequate data for appropriate statistical analysis with special attention to short- and long-term effects, local and regional effects, individual species and broad community impacts.
- Evaluate and review the test results.
- Adjust the monitoring program and/or practices according to requirement of the study area.

7.2.6.1 Ecosystem Protection

It is well established that some effects of mining can be predicted easily because they occur immediately, but other effects occur over a longer period and are more difficult to predict and measure. Ecosystem and its protection are one such long-term effect. The environmentally benign process (best practice mining) should include steps to identify these long-term effects and incorporate them in the management plan (EPA, 1995). Long-term impacts include the following:

- Vegetation changes caused by alteration in groundwater table
- Vegetation growth and health changes due to air pollution, dust, etc.
- Changes in the wildlife and bird migration pattern
- Extinction of some species of flora from the area
- Extinction of some faunal species from the area
- Ecological instability of post-mining landforms after mine closure

Since it is not practically feasible to monitor the entire ecological and biological systems, indicator species, processes, group or communities are usually selected. Their selection should be based on their efficacy or sensitiveness in the local system. Long-term and short-term plans and measures should be adopted for fragile hill areas.

7.2.6.2 Climate Change

Climate change studies by international agencies such as the United Nations have indicated that global warming has occurred mostly after the 1950s. Obviously, mountains, and in particular the Himalayan ecosystem, have an important contributory role to play in the various factors of global warming. Human population and developmental activities are responsible largely for it. Warming of the Earth has thus close relation with the hills, and climate change is presently being studied by scientists all over the world. Nearly 95% of Earth and atmospheric scientists are of the view that human intervention has led to the rise in the carbon dioxide level in the Earth's atmosphere, which is resulting in climate change. This CO_2-level rise is harmful to human health also. In order to protect the Himalayan ecosystem from the negative fallout of global warming and the greenhouse effect, attempts should be made to check (if not normalize) the CO_2 level so that at least further deterioration can be prevented. If the Earth keeps getting warmer at the current high rate, the impact on the monsoon pattern will become obvious, as winds causing monsoonal rains are affected due to the lofty hills.

We often hear that the 'Himalayan glaciers' are melting and receding faster than ever before. If the present rate continues, the likelihood of their disappearance may arise in even in the near future. The melting of snow in the glaciers at high altitude areas, glacial run-off, flash floods (Sharma et al., 2003), huge landslides, increase in earthquake frequency, etc., these are all directly or indirectly related to the climate change in the Himalayas. Climate change has led to rain rather than snowfall at higher altitudes of the Himalayas. It also accelerates the melting of glaciers.

The Intergovernmental Panel on Climate Change has already started the study of the climate changes in the Himalayas (IPCC, 2001), and no climate change policy or treaty will be complete without including the Himalayas and Himalayan communities. Since the lives of thousands of hill people are connected with climate change, which in turn is closely woven with environment and ecology of the region, it should not be left unattended. Participatory process for Himalayan communities to engage in the discussions on climate change, including disaster preparedness, must be kept in mind. In brief, though the objective of this book is not to focus on climate change in the Himalayas, an obvious climate change in the Himalayas is being observed. This has an effect on the plains ultimately.

7.3 Roadmap

The roadmap of mining and environmental practices for future should have the following points, which are based on the existing practices of mining in the Himalayas:

- With respect to the mining in the Himalayas, environmental protection should receive top priority and overexploitation should be stopped to meet the need and greed of the rising human demand.
- For environmental reasons, underground mining operations for mineral extraction should be preferred to surface mining.

- Limestone is the most abundant mineral of the Indian Himalayas and mined by surface mining methods only. Open-cast stripping of hills for large-, medium- and small-scale mining (with various degrees of mechanization) is feasible to fulfil the industrial requirement of this Himalayan region, but it should not be at the cost of environmental degradation.

- Manual small-scale mines occupy a significant portion of the mineral production in the Indian Himalayas, and these mines contribute about 40% of the total production. In other Himalayan countries such as Bhutan and Nepal, mining is largely done by semimechanized or manual operations of small and medium degree. Such practices with continual improvement should be followed in future.

- Mining in the small-scale private sector in the Himalayas is largely unplanned and non-systematic and therefore termed as 'unscientific'. Such exploitation practices do not have proper mechanisms to take care of environmental attributes in a scientific manner. This has also led to public protests and legal battles. Hence, 'scientific mining' should be promoted in future among small-scale private sector mines of the Himalayas.

- Sophisticated technical details for small-scale mine owners of the Himalayas are not easily available with respect to 'environment' and 'land data'. Therefore, land or environmental management approaches on the basis of some basic inputs such as soil depth, slope, existing land use, pH of land and its texture are desired.

- Because of financial constraints and illiteracy, complex technology has limited scope for successful application in the Himalayan ecosystem. Cost-effective technology for reclamation and restoration has wider applicability to potentially very fertile areas, such as valleys and nala sides only, but not in the barren lands of the hills.

- Incorrect or wrong historical practices should not be repeated in the future, and significant environmental repercussions on the Himalayan environment (e.g. mines in Doon Valley) should be repaired through learning.

- Significant modifications in geomorphology, soil characteristics, geohydrological regime, vegetation changes and water quality changes in the mining areas are caused as a result of various developmental works, not necessarily related to mining only. Such modifications directly or indirectly induce serious environmental hazards such as landslides, rock falls and land subsidence. These changes also restrain environment-oriented hill development. Their overall environmental impacts have at many places assumed serious dimensions, thereby demanding total scrapping of mining activities in the areas. In future, newer scientific approaches should be used for upgradation so that the accrued benefits can be harnessed by the hill people.

- Many a time, geological and geoenvironmental constraints arising out of the exploitation of complex Himalayan deposits are the reason for environmental damages, as no cost-effective alternatives are available. Such problems can be tackled easily by a scientific approach during operation or at the time of developmental planning.

These points can become guiding principles for mining in the Himalayas or a similar hill system even in other countries. On this basis, the following conclusion can be drawn, which we call the 'lessons learned'.

7.3.1 Lessons Learned

Environment-oriented development is the best solution to success for the overall mining operations in the Himalayas, which commences from exploration and ends with reclamation (Figure 7.1). To meet the social obligation of the Himalayan mining area development, a 'corpus fund' (collectively

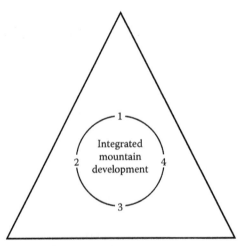

*1 = Sustainable development
*2 = Research
*3 = Information exchange
*4 = Sustainable management of environment
　　 and natural resources

Conceived from International Centre for Integrated
Mountain Development (ICIMOD), Nepal

FIGURE 7.1
Environment-oriented integrated hill development.

built up by the stakeholders over the operational stage of mine) in high-priority areas like education, drinking water, health care, sanitation, welfare of women, children, senior citizens and the disabled is desired. To create a green, supportive and conducive living environment (public buildings, roads, bridges, water supply systems, parks, irrigation, green energy, etc.) and for environmental conservation, this fund will be used either by the mine lessees or by the government authorities appropriately. In brief, this makes us to learn that in tough and rugged terrain conditions, extraction of minerals should be limited, selective and controlled and only green development should be promoted.

To keep the environmental degradation under control, the following points should be kept in mind:

- No single element of mining or environment management alone can by itself minimize the negative impacts of development intervention, for example mining.
- Shortcuts should be avoided as far as possible.
- An integrated strategy on watershed management should be the approach for planning.
- BMP should be adopted as far as possible.
- Very large-scale mechanization has limited scope on account of environmental sensitivity.

Research and precedence have indicated that a scientifically planned strategy should be the only approach for the excavation of minerals and preservation/protection of environment and the ecology in the sensitive Himalayan region.

Appendix 7.A: Land Management Strategy

The term *land management* may have many connotations. Therefore, it is necessary and appropriate to establish the meaning of the term. Here, it essentially relates to land disturbances by mining and relates to land capability and improvement thereof. Thus, land management means 'managing the transformed mining land to recovery, so as to bring it into a condition for ultimate use of agriculture or other uses'. It does not imply that the original conditions of the site existing before disturbance will be hitherto achieved but the progressive improvement from category VIII to category I or II land should be the objective for managing a mined land. Such practices thus enhance the transformation of wasteland to usable land. For various land categories, that is from I to VIII types, standard connotations will hold good.

Land Category	Management Strategy
Class I	1. Requires less to very less management. Suitable for agriculture.
	2. Immediate improvement of land by fertilizers and natural manures are not required urgently, but according to the species selected for planting, plant nutrients can be added.
	3. To improve yield in the long term, adopt new methods of farming with changed crops every 2–3 years. Leave the land barren for 1 year after every 3–4 years of cropping.
	4. Suggested bench terraces with H/V ratio as minimum 1 with 'Hydrem irrigation scheme'.
Class II	1. Requires less to very less management for quality improvement but disposal of excess water by properly planned method of conduit/drains with silt filtration arrangement is required.
	2. Nutrient loss of land due to water logging and its draining, together with soil erosion, requires immediate improvement of land by fertilizers and manure.
	3. To improve yield in the long term, adopt new methods of farming with changed crops every 2–3 years. Leave the land barren for 1 year after every 3–4 years of cropping.
	4. Suggested bench terraces with H:V ratio as minimum 1 with 'Hydrem irrigation scheme'.
	5. Suitable for agriculture.
Class III	1. Choice of plants is reduced and land requires conservation practices. Conservation measures and preliminary improvement of land by additives are required.
	2. Application of plants with high survival rates and high yields are recommended.
	3. Land is better suited for raising cultivated crops and cash crops.
	4. If the clay percentage is high, add the sand or coarse soil for improvement.
	5. Scope for mulching exist and may be required in case of used land. Decision has to be taken on a case-to-case basis.
Class IV	1. Improvement of land is required by terrace restoration in a sequence of shrub, grasses and trees.
	2. Proper designing of terraces and safe disposal of excess water is suggested.
	3. Organic manure in excess of normal is required for long-term restoration.
	4. Non-irrigated terraces are to be brought under the irrigated terraces category and land improvement as suggested is to be carried out.
	5. Changing pattern of vegetation is to be adopted for long-term improvement of land quality, and cosmetic plantation is the best to adopt in such land type.
	6. Class IV lands are marginally suitable for cropping and have poor production potential for cereal crops but can be better utilized by planting fruit and fodder trees spaced widely. Cash crops such as ginger, turmeric, beans, chilli and tomatoes can be sown along with the fruit trees to fill the crown cover of the trees and allow their cultivation.

(Continued)

Land Category	Management Strategy
Class V	1. Suitable for wildlife food and cover. Therefore, land management and improvement should be according to local conditions.
	2. Increases the organic content and humus content.
	3. Terracing is recommended.
	4. Use Hydrem irrigation scheme for raising the soil wetness if the climate condition permits.
	5. Intensive management is needed for the construction and management of inward sloping terraces.
Class VI	1. Development as pasture land or grazing land is to be preferred in such cases; therefore, first, phase management is needed. Later on, the land may be left for natural improvement.
	2. Terrace restoration is done through grass and planting of wild and top-feed trees.
	3. Increase in organic content/humus content is suggested.
	4. Slope grading is suggested for effective increase in land area for development.
	5. Mixed grass and legume variety of plantation on this land is desirable.
Class VII	1. Large quantity of organic matter and plant nutrient to the soil is needed. Thus, land management is of very severe order.
	2. Natural grass cover or planting of brushes and shrubs is preferred as management strategy to reduce the run-off and soil losses.
	3. If soil pH is acidic, mixing of ground limestone (90% which can pass through a 10 mesh screen and 50% through a 60 mesh screen) with the soil to a depth of at least 6 in. is recommended. Agricultural grade limestone is to be preferred.
	4. Improvement in a phased manner for long-term land improvement is suggested.
	5. Slope grading is suggested to check soil erosion and water conservation.
	6. Adequate arrangements for increasing water retention capacity or wetness of land/soil are required.
Class VIII	1. This land class is a non-cultivable, degraded land totally unsuitable for production purposes. Therefore, other methods, such as conversion to a pond for aquaculture and conversion to a polo ground or playground for recreation, are the best form of management. Very, very severe management is required to protect and restore vegetation cover and control erosion. Therefore, land treatment suggested should be based on the site condition.
	2. 'Progressive improvement' by scientific investigations is the best way of management.
	3. Natural restoration during initial stages is recommended.

8

Parvatadhiraj: The Himalaya

Before I start writing this last chapter, let me answer one pertinent question which might arise in an inquisitive mind while looking at this page – why is this chapter so named and why is it needed?

In Hindu mythology, Himalaya is considered as 'Parvatadhiraj', meaning 'the king of mountains'. In ancient Hindu epics and literature, references to 'Himalaya Parvat' (पर्वत) are aplenty, which is considered the abode of God (the word 'Parvat' means mountain in Hindi). 'The Himalaya' is one of the most sacred (holiest) places on Earth, where nature shows all her beauty and diversity in the best form. The serene environment of the entire region makes it ideal for worshipping, meditation and *nirvana* by Hindu *rishis, sadhus, mahatamas, pundits* and meditators. Even today, the Himalayas are the preferred destination for all humans around the globe. Sir Edmund Hillary (the first European mountaineer to climb Mount Everest, the highest peak of the Himalayas) and several other mountaineers, who have seen it closely, have narrated and described the bountifulness of the Himalayas as unparalleled. This is the reason why the chapter is titled 'Parvatadhiraj'.

Figure 8.1 shows the Hindu deity 'Lord Ganesha' (the first god worshipped in all Hindu spiritual prayers) with a view to concluding the book with a prayer to god and seek the Almighty help to protect the sensitive Himalayas from natural disasters and calamities, and thereby all of us. Since these natural disturbances are triggered by humans but are beyond their control, I believe the naming of the chapter (Parvatadhiraj) and placement of the picture of *Lord Ganesha* are correct and appropriate.

8.1 Responsible Mining

The twenty-first century is witnessing noticeable changes in the industrial sector, with mining as no exception. Mining and the environment have to move in unison for societal benefits, and without this, mining in sensitive areas will have a bleak future. Himalayan mining should be 'responsible mining'. It requires sustainable mining with the best management practices. It should be the dictum for all mining companies operating in the Himalayas, be they are large or small.

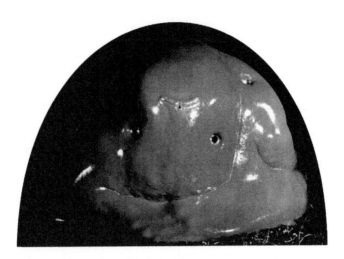

FIGURE 8.1
Lord Ganesha.

In fragile hill regions, a mining company should operate in such a way that the activities leave fewer carbon footprints. Its benefits can be taken in those areas where mining operations exist within 10 km of 'protected areas', such as biological hotspots, wildlife sanctuaries etc. A large mining company must initiate and prepare a 'wildlife conservation plan' and a 'biodiversity action plan' for the management of biodiversity and wildlife in various Himalayan hill regions and add to the national database.

Besides adopting modern technologies in the drilling and blasting process, they should also cover transportation, communication and mechanization of various other unit operations of mining. Himalayan mines should also develop raw material optimization through the latest computer applications, for example quarry scheduling optimization. Mine planning using modern software has plenty of scope in the Himalayan mines.

Proper use of computers and information technology in various fields of engineering and management can reduce the carbon footprints and make Himalayan mining 'responsible mining' now and in the future.

8.2 Epilogue

Three common designations of mountains are of particular interest for research studies: namely 'mountain', 'mountain range' and 'highland'. Himalaya is a mountain and not a *mountain range*. It is a landform displaying a conspicuous relief with moderate to high elevation and a discernible, small summit area. It should not be confused with a highland either, as 'highland' is defined as an extensive landform with a mixture of mountains, hills and plateaus.

To finally conclude, we may summarize mining in the Himalayas as follows:

Mining in the Himalayas is a contentious issue and raises several questions. In Chapter 3 of this book, where we spoke about the 'eco-perspective of Himalayan mining', some burning questions were answered technically. These answers give direction to many contentious issues. In brief, by reading this book, a reader will get pertinent answers and learn more about the Himalayas and Himalayan mining, thereby enriching one's knowledge on the Himalayan ecosystem.

1. Though the Himalayas are a storehouse of minerals and have immense possibilities for future also, considering its natural beauty, environmental sensitiveness and proneness to natural calamities such as earthquakes and massive slope failures, it is safer and better to keep mining of minerals in the region at a low level and leave the mineral treasures for future generations.

2. Problems and prospects of mining and mineral sector development are neither simple nor easy. Protection of the environment and ecology during mining and post mining should be on top of the development agenda.

3. The policy for the development should be result oriented, time bound, transparent and eco-friendly. The developmental policy of the mineral-bearing areas of the Himalayas must consider and clearly emphasize the priorities with respect to the existing mineral, industrial and forest policies. This will help in achieving sustainable development.

4. Commercial exploitation of mineral resources must prompt us to think at what cost we will get them. If we continue the exploitation of minerals, we must consider whether we are damaging the environment beyond repair or we are leaving some for future generations.

5. If mining operations are being continued by adopting an integrated strategy, with best practice mining and an environment-oriented development, this approach can be considered acceptable to the maximum.

6. Last but not least, we have to learn a lot from the Parvatadhiraj Himalaya in future, as it is the most active field laboratory for geologists and the mineral industry professionals (the Himalayas are rising!).

Consider the quote of Sir Edmond Hillary, a New Zealander: 'Environmental problems are essentially social problems they begin with man as the cause and end with man as victim'. This is cent percent true with the fragile Himalayas, which we discussed in this book. Himalayan mining and environmental problems must be dealt with a social perspective so that a man does not become a victim.

Finally, by writing this concluding chapter, I aim to achieve two things: (1) be able to pay tribute to the king of mountains (Parvatadhiraj Himalaya), fulfill a tradition and make a dedication and (2) be able to make my attempt to bridge the knowledge gap among science, engineering and the environment with respect to the mighty Himalayas.

Himalayan Mining: From Real to Reel (Memories Down the Lane)

FIGURE 1
A view of hilltop mining in the Indian Himalayas.

FIGURE 2
A view of a barren hill when there was no mining at Kashlog (picture taken in 1994). (Picture Courtsey of Ambuja Cement, Darlaghat, Himachal Pradesh, India.)

FIGURE 3
A view of slope failure in the Himalayas due to natural reasons: Malti landslide.

FIGURE 4
Two old pictures of Gagal hill for the mining of limestone for cement making.

FIGURE 5
Collage of a closed mine in the Himalayas (picture taken in 1996).

FIGURE 6
Starting phase of a mine in the Himalayas, 1994–1995.

FIGURE 7
Decommissioning phase/reclamation work, 1985.

FIGURE 8
Restoration of mined-out area/reclaimed land area of a mine in the Himalayas (1996–1997).

FIGURE 9
A mine and a nearby village in the Indian Himalayas.

References

Abbassi, T. and Abbassi, S.A. (2012). *Water Quality Indices*. Elsevier Publication, Oxford, U.K. and Amsterdam, The Netherlands, p. 384.

Anon. (1973). Minerals for industrial use. Government of Meghalaya, Directorate of Mineral Resources, Shillong, India, p. 38.

Anon. (1976). Mineral deposits of Himalaya – A brief review of present status and future possibilities. *Proceedings of Himalayan Geology Seminar*, Geological Survey of India (GSI) Publication, New Delhi, India, p. 19.

Auden, J.B. (1934). *Geology of the Krol Belt*, in Krishnan, M.S. Record 67(4).

Banerjee, P.K. (1993). Environmental problems related to mining in hilly region. *Proceedings of Seminar on Environmental Policy Issues in Mineral Industry, MGMI*, Calcutta, India, pp. 89–95.

Basu, S.R. and Saxena, A.K. (1984). Techno-economic feasibility of transporting limestone in slurry form to feed a cement plant at the foothills of the Himalaya. *International Journal of Bulk Solid Handling*, 4(3), 583–588.

Bhandari, R.K. (1987). Slope instability in the fragile Himalaya and strategy for development. *Ninth Indian Geotechnical Society (IGS) Annual Lecture*, IIT, New Delhi, 16 December 1987, pp. 1–87.

Bhargava, D.S. (1983). Use of a water quality index for river classification and zoning of Ganga River. *Environmental Pollution (Series B)*, 6, 51–67.

Bhargava, O.N. (1976). Geology of the Krol Belt and associated formations: A reappraisal. *GSI Memoir*, 106(1), 167–234.

Burn, D.H. and Yuliatani, S. (2001). Waste load allocation using genetic algorithm. *Journal of Water Resource Planning Management*, 127(2), 121–129.

Canter, L.W. (1996). *Environmental Impact Assessment*. McGraw Hill, New York, p. 655.

Chakraborty, S.L. (1997). Cluster mining: A tested concept for small mines. In *Small/Medium Scale Mining: A Global Perspective*, A.K. Ghose (ed.), National Institute of Small Mines (NISM), Oxford, U.K., and Calcutta and IBH Co Pvt Limited, New Delhi, pp. 59–64.

Chakraborty, S.L. (2001). Artisinal and small scale mining in India. Mining, Minerals and Sustainable Development (MMSD), A project by International Institute for Environment and Development (IIED), London and World Business Council for Sustainable Development (WBCSD), Published as United Nations Documents relating to Mining Industry by UNEP and UNCTAD, October 2001, p. 81.

Chakraborty, E., Kumar, G.V.V., and Kumar D. (2014). Detection of hazard prone areas in the upper Himalayan region in GIS environment. *International Journal of Research in Engineering and Technology (IJRET)*, 3(16), 141–149 (special issue).

CPCB. (1990). Environmental standards for prevention and control of pollution. Central Pollution Control Board, Government of India, New Delhi, India.

CSIR-CIMFR. (2007). Technical guidelines for controlled blasting (for opencast and underground mines, p. 53). CSIR-Central Institute of Mining and Fuel Research (CIMFR), Dhanbad, India.

CSIR-CIMFR. (2012). Assessment of technical feasibility of mechanical mining for (lower pit) Alsindi limestone deposit of Himalaya, District - Mandi, Himachal Pradesh. Technical Report No. CNP/N/2781/2010-11. CSIR-Central Institute of Mining and Fuel Research (CIMFR), Dhanbad, India, p. 66.

CSIR-CIMFR. (2014). Slope stability studies for proposed lower pit at Alsindi limestone deposit of Lafarge India Private Limited. Technical Report No. CNP/N/2782/2010-11. CSIR-Central Institute of Mining and Fuel Research (CIMFR), Dhanbad, India, p. 53.

Dar, K.K. (1968). Metallogeny in the Himalayas. *Indian Geological Congress, 23rd Session*, Vol. 7, pp. 35–42. In Nair, N.G.K. and Mithal, R.S., 1976.

Daylami, A.A., Shamsi, A., and Niksokhan, M.H. (2010). Model for waste load allocation in rivers: A cooperative approach. *American–Eurasian Journal of Agriculture and Environment Sciences*, 8(6), 626–636.

Desai, M. (2014). *Darjeeling the Queen of Hills: Geo Environmental Perception*. K.P. Bagchi & Company, Kolkata, India, p. 206.

Dhawan, G. (2013). Managing geological problems during construction of hydropower tunnels in India with respect to Himalayas. In *Proceedings of the Global Views of Engineering Geology and the Environment*, F. Wu and S. Qi (eds.). Taylor & Francis, pp. 231–239.

Dhurvanarayan, V.V., Sastry, G., and Patnaik, U.S. (1990). *Watershed Management*. Indian Council of Agricultural Research (ICAR), New Delhi, India, p. 176.

Ding-Quan, L., Li-Zhong, L., Guang, M., and Guo-Lin, L. (1989). Some aspects of mining in high mountain areas above snow line in China. In *Mining in Arctic*, S.N. Bandopadhya and A.A. Skudrzyk (eds.). Balkema, Rotterdam, the Netherlands, pp. 217–225.

DMG (Department of Mines and Geology). (1991). Present status of mineral resources development in Nepal. In *Small Scale Mining: A Global Overview*, A.K. Ghose (ed.). Oxford IBH Company Private Ltd., New Delhi, India, pp. 241–257.

EPA. (1995). Best practice environmental management in mining – Various modules. Environmental Protection Agency (EPA), Perth, Western Australia, Australia.

EPA. (2002). Overview of best practice environmental management in mining. Environmental Protection Agency (EPA), Perth, Western Australia, Australia.

Garg, J.K. (1990). Impact of mining activities and super thermal power stations on environment. Project Report No. RSAM/SAC/RSAG/ENVN/PR/8/90. Space Application Centre, Ahmadabad, India.

Ghose, A.K. (1993). India's minerals future – Mega trends for the next millennium. *First Foundation Lecture of Indian Geological Congress*, Roorkee, India, 11 December 1993.

Ghose, M.K. (2003a). Promoting cleaner production in the Indian small-scale mining with special emphasis on environmental management. *Journal of Cleaner Production*, 11, 159–165.

Ghose, M.K. (2003b). Promoting cleaner production in the Indian small-scale mining industry. *Journal of Cleaner Production*, 11, 167–174.

Ghose, M.K. (2004). Emission factor for the quantification of dust in Indian coal mines. *Journal of Industrial Research*, 63, 763–768.

Goel, P.K. and Sharma, K.P. (1996). *Environmental Guidelines and Standards in India*. Techno-Science Publications, Jaipur, India, p. 318.

GOI. (1991). *Industrial Policy*. Government of India, New Delhi, India.

Guild, P.W. (1971). Metallogeny: A key to exploration. *Mining Engineering*, 23, 69–72.

Gupta, A.K. (1992). An integrated approach for development of hill areas study using satellite and collateral data. National Natural Resources Management System (NNRMS), Department of Space Project Report, Bangalore, India, p. 57.

Gupta, R.K. (1983). Land use regulation for flood control and watershed management in the Himalayas. *Indian Journal of Soil Conservation*, 11(1), 10–29.

Gupta, R.K. and Arora, Y.K. (1985). Technology for the rejuvenation of degraded lands in the Himalaya. In *Environmental Regeneration in Himalaya: Concepts and Strategies*, J.S. Singh (eds.). Central Himalayan Environment Association, Nainital, India, pp. 169–186.

Hilson, G. (2003). Environmental management in the small-scale mining industry. *Journal of Cleaner Production*, 11, 91–93 (special issue editorial).

IBM. (1982). Monographs on limestone and dolomite. Technical Consultancy, Mining Research and Publication Division, Indian Bureau of Mines (Government of India), Nagpur, India, pp. 5-100–5-103.

IBM. (1994). Environment aspects of mining area. Bulletin No. 27. Training, Mining Research and Publication Division, Indian Bureau of Mines, Nagpur, Publication of Ministry of Mines, Indian Bureau of Mines (Government of India), Nagpur, India.

IBM. (2009). Indian minerals yearbook 2009. Indian Bureau of Mines (Government of India), Ministry of Mines, Nagpur, India.

IBM. (2014a). Indian minerals yearbook – 2012. Part III: Mineral reviews, 51st edn. Indian Bureau of Mines (Government of India), Ministry of Mines, Nagpur, India.

IBM. (2014b). Indian minerals yearbook – 2012. Part I: Sikkim state reviews, 51st edn. Indian Bureau of Mines (Government of India), Ministry of Mines, Nagpur, India.

IBM. (2015). Indian minerals yearbook 2013, 52nd edn. (advance release). Part 1: General reviews. Ministry of Mines, Indian Bureau of Mines, Government of India Publication, Nagpur, India, pp. 1-2–1-24.

ICIMOD. (1983). Mountain development – Challenges and opportunities. *Proceedings of the First International Symposium*, Kathmandu, India (Country statements and selected papers), p. 122.

ICIMOD. (1995). International Centre for Integrated Mountain Development: An Overview. ICIMOD, Kathmandu, Nepal (In-house publication of ICIMOD), p. 36.

Inamdar, A.B., Venkatraman, G., and Natrajan, C. (1992). Landuse-landcover studies through remote sensing of Kashlog project. Centre of Studies in Resource Engineering, Indian Institute of Technology, Bombay, India, p. 17.

IPCC. (2001). The Intergovernmental Panel on Climate Change (IPCC). Third Assessment Report, Vol. I (The scientific basis), Vol. II (Impacts, adaptation and vulnerability), and Vol. III (Mitigation). Cambridge University Press, Cambridge, U.K.

Jayan, T.V. (2004). Mining the Himalaya. *Down to Earth*, 15 January 2004.

Juwarkar, A.A. and Jambhulkar, H.P. (2008). Phyto-remediation of coal mine spoil dump through integrated biotechnological approach. *BioResource Technology*, 99, 4732–4741.

Juwarkar, A.A., Singh, L., Singh, S.K., Jambhulkar, H.P., Thawale, P.R., and Kanfade H. (2015). Natural vs. reclaimed forests – A case study of successional change, reclamation technique and phyto-diversity. *International Journal of Mining, Reclamation and Environment*, 29 (06), 476–498.

Juwarkar, A.A., Yadav, S.K., and Thawale, P.R. (2010). Bio-technological approach for ecosystem restoration of mine spoil dump in India. *International Journal of Environment and Pollution*, 43, 251–263.

Juwarkar, A.A., Yadav, S.K., Thawale, P.R., Kumar, P., Singh, S.K., and Chakrabarti, T. (2009). Developmental strategies for sustainable ecosystem on mine spoil dumps: A case of study. *Environment Monitoring and Assessment*, 157, 471–481.

Keller, E.A. (2011). *Environmental Geology*, 11th edn. Pearson Prentice Hall, Upper Saddle River, NJ, p. 596.

Kennedy, A. (1993). Limestone transportation by capsule pipeline. *Mining Magazine*, December, pp. 294–298.

Kleynhans, C.J. (1996). A qualitative procedure for the assessment of the habitat integrity status of the Luvuvhu River limpoposystem, South Africa. *Journal of Aquatic Ecosystem Health*, 5, 41–54.

Krishna, J.S.R., Rambabu, K., and Rambabu, C. (1995). Monitoring, correlations and water quality index of a well waters of Reddiguden Mandal. *Indian Journal of Environmental Protection*, 15(12), 914–919.

Krishnan, M.S. (1982). *Geology of India and Burma*. CBS, New Delhi, India, p. 536.

Kutschera, S. (1984). Planning aspects for the application of continuous transport systems in hard rock open pit mines. *International Journal of Bulk Solid Handling*, 4(3), 609–613.

Malhotra, C.L. (1997). Applicability of solution mining to Drang salt deposit at Mandi. *Proceedings of the National Seminar on Eco-Friendly Mining in Hilly Region and Its Socio Economic Impacts (HILMIN'97)*, June 1997, Mining Engineers Association of India (MEAI), Himalayan Chapter, Shimla, India, pp. 151–154.

Mathur, V.S. (1994). Control measures of water resources and its quality while mining in Himalaya. *Souvenir – Mines Environment and Mineral Conservation Week'94*. Organized by IBM, Dehradun and Mine Lessee of Dehradun Region, Dehradun, India, pp. 1–3.

Mauriya, V.K., Yadav, P.K., and Angra, V.K. (2010). Challenges and strategies for tunneling in the Himalayan region. *Indian Geotechnical Conference'2010, GEO Trendz*, 16–18 December 2010. Indian Geotechnical Society (IGS) Mumbai Chapter and Indian Institute of Technology (IIT), Bombay, pp. 93–96.

Maynard, D. and Walmsley, M. (1981). Terrain and soil assessment in support of planning for metal mine development. *CIM Bulletin*, 74(834), 66–71.

MMSD. (2001). Artisinal and small scale mining in India. Mining, Minerals and Sustainable Development (MMSD): A project by International Institute for Environment and Development (IIED), London and World Business Council for Sustainable Development (WBCSD), Published as United Nations Documents relating to Mining Industry by UNEP and UNCTAD, October 2001, p. 81.

Moddie, A.D. (1980). Environment oriented hill development. In *Studies in Himalayan Ecology and Development Strategies*, T. Singh (ed.). English Book Store, New Delhi, India, pp. 193–198.

MOEF. (1990). Parameters for determining ecologically fragility. Ministry of Environment, Forest and Wild Life (MOEF), A Government of India Publication, New Delhi, India, pp. 1–14.

MOEF. (1992). The policy statement for abatement of pollution – 1992. Ministry of Environment and Forest, Government of India, New Delhi, India, p. 9.

Mohnot, J.K. and Dube, A.K. (1995). Scientific concepts for environment friendly mining technology for fragile ground conditions of Himalaya. *The Indian Mining and Engineering Journal*, 34(6), 23–27.

Mukherjee, S., Carosi, R., Beek, Peter van der, Mukherjee, B.K., and Robinson, D.M. (2015). Tectonics of the Himalaya: An introduction, Research Publication of Lyell collection, Geological Society of London – Online publication, pp. 1–3. http://sp.lyellcollection.org.

Mukherjee, S., Mukherjee, B.K., and Thiede, R.C. (2013). Geosciences of the Himalaya-Karakoram-Tibet Orogen, Editorial. *Indian Journal of Earth Sciences* (*Geolo Rundsch*), 102, 1757–1758.

Mukherjee, S.K. (1983). Some aspects of Himalayan geology with special reference to strategy for future mineral exploration. Centennial Lecture. Wadia Institute of Himalayan Geology, Dehradun, India, pp. 1–31.

Nagarajan, R., Banerjee, S.P., and Roy, A. (1994). Remotely sensed data analysis of limestone mines and its environ along Giri River, Sirmour District, H.P. *Second National Seminar on Minerals and Ecology*. Dhanbad, Oxford & IBH Company Ltd., New Delhi, India, pp. 21–27.

Nair, N.G.K. and Mithal, R.S. (1976). The Alps and the Himalaya: A comparison of their metallogeny. *Journal of Himalayan Geology*, 6, 287–302.

Noetstaller, R. (1994). Small scale mining: Practices, policies, perspectives. In *Small Scale Mining: A Global Overview*, A.K. Ghose (ed.). Oxford IBH & Company, New Delhi, India, pp. 3–10.

Ott, W.R. (1978). *Environmental Indices: Theory and Practice*. Ann Arbor Science Publisher Inc., Ann Arbor, MI, p. 371.

Pachauri, A.K. (1992). Plate tectonics and metallogeny in the Himalaya. In *Himalayan Orogen and Global Tectonics*, A.K. Sinha (ed.). International Lithosphere Programme Publication No. 197, Routledge, pp. 267–288.

Paithankar, A.G. (1993). Large scale hill top mining with concern for the environment. In *Proceedings of the Innovative Mine Design for the 21st Century*, W.F. Barvden and J.F. Archibald (eds.). Balkema, Rotterdam, the Netherlands, pp. 331–335.

Pandian, S. and Prasad, R. (2014). Study on non-exhaust particulate matter emission factors in urban and mining areas. *MinENVIS News Letter*, 83, 1–3.

Prushti, B.K. (1996). Socio economic profile of Bhowra area of BCCL. MTech dissertation (unpublished). Centre for Mine Environment (CME), Indian School of Mines, Dhanbad, India.

Rai, K.L. (1993). Geological and geo-environmental aspects of mineral resources development in lesser Himalayan tracts of Sikkim and U.P. Himalaya, India. *Proceedings of the Asian Mining – 1993*, Calcutta, India.

Roder, W. (2000). Management of natural resources in Bhutan, Chapter 8, pp. 149–165. http://lib.icimod.org/record/21051/files/c_attachment_54_359.pdf.

Roy, S.C. (1993). A geo-environmental appraisal of impacts of mining in Sataun Dadahu area of Sirmour District, Himachal Pradesh. *Indian Minerals*, 47(4), 329–334.

Saklani, P.S. (1978). A review of the stratigraphy of lesser Himalaya of U.P. and H.P. In *Tectonic Geology of the Himalaya*, B.N. Raina (ed.) Today & Tomorrow's Printers and Publishers, New Delhi, India, pp. 79–112.

Saxena, N.C. (1995). Environmental management plan (EMP) preparation for mining projects – An approach. *The Indian Mining and Engineering Journal*, 34(5), 35–43.

Saxena, N.C. and Banik, P.K. (1996). Land use planning. *Proceedings of One Week Executive Development Program*, Centre of Mining Environment, Indian School of Mines, Dhanbad, India, 22–26 July 1996.

Saxena, N.C., Singh, G., and Ghosh, R. (2002). *Environmental Management in Mining Areas*. Scientific Publishers (India), Jodhpur, India, p. 410.

Sen, P., Ghose, A.K., and Mozumdar, B.K. (1991). Emplacement design for a hilly iron ore mining in India. *Journal of Institution of Engineers (India)*, 72(Part MN1), 1–3 (Mining Engineering Division).

Sharma, A.K., Agnihotri, R.N., and Sud, J.K. (1997). Environmental awareness among small scale mines – A case study of Baldhwa limestone mine of M/s Jai Singh Thakur and Sons, District Sirmour, Himachal Pradesh. *Proceedings of the National Seminar on Eco-Friendly Mining in Hilly Region and Its Socio-Economic Impacts (HILMIN'97)*, MEAI Himalayan Chapter, Shimla, India, June 1997, pp. 276–285.

Sharma, V., Mishra, V.D., and Joshi, P.K. (2013). Implications of climate change on stream flow of a snow-fed river system of the northwest Himalaya. *Journal of Mountain Science*, 10(4), 574–587.

Shapkota, P. (1991). Small scale mining in Bhutan: A perspective. In *Small Scale Mining: A Global Overview*, A.K. Ghose (ed.). Oxford IBH Company Private Ltd., New Delhi, India, pp. 221–239.

Silitoe, R.N. (1969). Speculation on Himalayan metallogeny based on evidences from Pakistan, Geo-dynamics of Pakistan, Geological Survey, Quetta, Pakistan, pp. 1–31. In Mukherjee, S.K., 1983.

Singh, S.K. (1996). Use of sequential blasting machine for deep hole blasting at Gagal Works. *Souvenir Volume of Mines Environment and Mineral Conservation Week*, 1995–96. Organised by IBM, Dehradun, publisher-Mine Lessee of Dehradun Region, Dehradun, India, pp. 67–68.

Sol, V.M., Lammers, P.E.M., Harry, A., Joop, D.B., and Jan, F.F. (1995). Integrated environmental index for application in land-use zoning. *Environmental Management*, 19(3), 457–467.

Soni, A.K. (1994a). Impact of mining on regional development: Case history of the Himalayan foot hills. In *Monograph on Role of Mining in the Development of the Backward Regions of India*, A.M. Hussain (ed.). Department of Science and Technology, Ministry of Science and Technology, Indian Bureau of Mines, Nagpur (A project document of Government of India), pp. 267–293.

Soni, A.K. (1994b). An appraisal of environmental problems of mining in Himalaya with possible remedies. *International Conference on Impact of Mining on the Environment: Problems and Solutions*. VRCE, Nagpur, India, pp. 253–266.

Soni, A.K. (1997). Integrated strategy for development and exploitation of mineral resources of ecologically fragile area. PhD thesis (unpublished), Indian School of Mines (ISM), Dhanbad, India, p. 238.

Soni, A.K. (2002). Land management based on land capability classification approach with reference to mining land of Himalayas. In *Environmental Pollution Research*, D.B. Tripathi and B.B. Dhar (eds.). APH Publishing Corporations, New Delhi, India, Chapter 10, pp. 111–134.

Soni, A.K. (2003). Utilization of slate mine waste. *Journal of Indian Association of Environmental Management (IAEM)* 30(2), 102–106.

Soni, A.K. and Dube, A.K. (1995). Slate mining in Kangra Valley: A pilot study embodying environmental concerns. *Proceedings of the First World Mining Environment Congress*, New Delhi, India, pp. 733–748.

Soni, A.K. and Dube, A.K. (2000). Environment oriented development of mining areas in Himalaya. *Proceedings of the International Conference on Geo Environment Reclamation*, Nagpur, India, November 2000, pp. 195–201.

Soni, A.K. and Kiran, S. (2012). The road to zero harm: Safety management and safety engineering in context to underground mines. *Workshop on Safety Management in Mines*, Nagpur, India, pp. 1–10.

Soni, A.K. and Loveson, V.J. (2003). Land damage assessment – A case study. *Journal of Indian Society of Remote Sensing*, Dehradun, 31(3), 175–186.

Sundarajan, M., Chakraboraty, M.K., Gupta, J.P., Saxena, N.C., and Dhar, B.B. (1994). Mathematical model for water quality impact assessment and its computer application in coal mine water. *Proceedings of the International Symposium on Impact of Mining on the Environment: Problems and Solutions*, Nagpur, India, pp. 91–101.

Tejal, P., Udit, J., and Payal, Z. (2014). Rail link project – A case study on Jammu-Udhampur-Srinagar-Baramulla. *International Journal of Current Research and Academic Review*, 2(2), 167–172.

Thakur, B. (ed.). (2008). Environmental degradation and possible solution for restoring the land: A case study of magnesite mining in the Indian central Himalaya. In: *Perspectives in Resource Management in Developing Countries*, Vol. 3: *Ecological Degradation of Land*, Concept's International Series in Geography-5. Concept Publishing Company, New Delhi, India, Chapter 14, pp. 244–260.

Thakur, D.N. and Kumar, B. (1993). Reclamation of steeply sloping mine waste dumps at high altitudes. *Proceedings of the Asian Mining*, Calcutta, India, pp. 105–116.

Thakur, D.N., Kumar, B., and Tiwary, B.K. (1992). Environmental management of mine wastes at high altitude. *Proceedings of the Fourth National Seminar on Surface Mining*. Indian School of Mines, Dhanbad, India, pp. 4.3.1–4.3.13.

Tripathy, D.P. and Reddy, K.G.R. (2015). Eco-friendly practices for sustainable development of mining industry. *Mining Engineering Journal*, 16(12), 23–29 (Mining Engineers Association of India, Hyderabad).

UN. (1993). Guidelines for the development of small/medium-scale mining: Selected papers. In MMSD, 2001. Presented at the *UN Inter Regional Seminar*, Harare, Zimbabwe, 15–19 February 1993.

UNEP. (1997). Mining in fragile area: A case study of slate mining. In *UNEP Training Manual on Environmental Management of Mining Operation*, First Indian Edition, B.B. Dhar (ed.). Oxford & IBH Publishing Company, New Delhi, India, pp. 6.28–6.36.

UNEP. (2014). Sections 13.1–13.24, Managing fragile ecosystem: Sustainable mountain development. Available at: http://www.unep.org/Documents.Multilingual/Default.asp?DocumentID=52 & ArticleID=61& l=en. Accessed on 26 July 2014.

USEPA. (2001). Coal re-mining best management practices guidance manual. U.S. Environmental Protection Agency, Washington, DC.

Valdiya, K.S. (1984). *Aspects of Tectonics*. Tata McGraw Hill Publication, New York, p. 342.

WB. (1995). *Artisinal Mining. Proceedings of the World Bank (WB) Round Table Conference*, Washington, DC, 17–19 May 1995. In MMSD 2001.

Yin, A. (2006). Cenozoic tectonic evolution of the Himalayan orogen as constrained by along-strike variation of structural geometry, exhumation history, and foreland sedimentation. *Earth-Science Reviews*, 76(1–2), 1–131.

World Wide Web

http://cpcb.nic.in: Indian laws and documents related to air and water pollution.

http://mines.nic.in: National Mineral Policy, Mines and Minerals (Regulation and Development) Act-1956; Mines Act-1952; Mineral Conservation and Development Rules (MCDR).

http://www.CSIR.co.za: River health indicators and indices.

http://www.ENVIS.Sikkim India: Mining in Sikkim state.

http://www.dossantosintl.com: High angled conveying.

http://www.dmg nepal.gov.np: Directorate of Mining and Geology, Government of Nepal.

http://www.moef.nic.in: Environment Protection Act (EPA), 1986 and National Forest Policy, 1988.

http://www.moef.nic.in/legis/eia: Indian environmental legislation, EIA, etc.

http://www.siemag-tecberg.com: Slope hoisting system.

Suggested Reading

Books

Bahadur, J. (2003). *Indian Himalayas: An Integrated View* (A Book on Agricultural Ecology). Vigyan Prasar, New Delhi, India, p. 279.

Chadha, S.K. (1989). *Himalaya Ecology*. Ashish Publishing House, New Delhi, India, p. 168.

Chaudhri, A.B. (1992). *Himalaya Ecology and Environment*. Ashish Publishing House, New Delhi, India, p. 596.

Desai, M. (2014). *Darjeeling the Queen of Hills: Geo Environmental Perception*. K.P. Bagchi & Company, Kolkata, India, p. 206.

Dikshit, K.R. and Dikshit, J.K. (2014). *North-East India: Land, People and Economy* (in V Parts). Springer, Berlin, Germany, p. 800. (Electronic Version).

Gerrard, J. (1990). *Mountain Environments: An Examination of the Physical Geography of Mountains*. Belhaven Press, London, U.K., p. 317.

Gujral, G.S. and Sharma, V. (eds.). (1996). *Changing Perspective of Biodiversity Status in the Himalaya*. British Council Division, British High Commission, New Delhi, India, p. 186.

Gupta, R.K. (1980). Alternative strategies for rural development in Garhwal Himalaya. In *Studies in Himalayan Ecology and Developmental Strategies*, T. Singh (ed.). English Book Store, New Delhi, India, pp. 218–233.

Ives, J.D. and Messerli, B. (1997). *The Himalayan Dilemma: Reconciling Development and Conservation*. Routledge, New York, p. 324.

Jodha, N.S. (1995). *Sustainable Development in Fragile Environments: An Operation Framework for Arid, Semi Arid and Mountain Areas*. Centre for Environmental Education (CEE), Ahmadabad, India, p. 122.

Joshi, S.C., Bhattacharya, G. (eds.); Pangtey, Y.P.S., Joshi, D.R., and Dani, D.D. (co-eds.). (1988). *Mining and Environment in India*. Himalayan Research Group, Nainital, India, p. 467.

Kishore, V. and Gupta, R.K. (1981). *Socio Economic Studies for Land Use Planning and Eco-Development in Garhwal Himalaya*. INTACH Publication, New Delhi, India, p. 89.

Lall, J.S. (ed.). (1981). *The Himalaya: Aspects of Change*. Oxford University Press, India International Centre, Delhi, India, p. 481.

Mark, T. (2015). *Mining in Ecologically Sensitive Landscapes*. CRC Press/Balkema, The Netherlands, p. 275.

Messerli, B. and Ives, J.D. (eds.). (1997). *Mountains of the World*. CRC Press, Boca Raton, FL, p. 510.

Nishizawa, T. and Uitto, J.I. (1995). *The Fragile Tropics of Latin America: Sustainable Management of Changing Environments*. United Nations University Press, Tokyo, Japan, p. 325.

Pirazizy, A.A. (1993). *Mountain Environment: Understanding the Change*. Ashish Publishing House, New Delhi, India, p. 194.

Rawat, M.S.S. (1993). *Himalaya, a Regional Perspective: Resources, Environment and Development*. Daya Publishing House, Delhi, India, p. 202.

Nibanupudi, H.K. and Shaw, R. (2014). *Mountain Hazards and Disaster Risk Reduction*. Springer Publication, Tokyo, Japan, p. 284.

Singh, J.S. (ed.). (1985). *Environmental Regeneration in Himalaya Concepts and Strategies*. Central Himalayan Environment Association Publication, Nainital, India, p. 468.

Singh, T.V. and Kaur, J. (eds.). (1980). *Studies in Himalayan Ecology and Development Strategies*. English Book Store, New Delhi, India, p. 286.

Stahr, A. and Langenscheidt, E. (2015). *Landforms of High Mountains*. Springer, Berlin, Germany, p. 158.

Tiwari, D.N. (1994). *Himalayan Ecosystem*. International Book Distributors, Dehradun, India, p. 355.

Valdiya, K.S. (1980). *Stratigraphy and Correlation in Lesser Himalaya*. Tata McGraw Hill Publication, New Delhi, India, p. 330.

Valdiya, K.S. (1984). *Geology and Natural Environments of Nainital Hills, Kumaon Himalaya*. Gyanoday Prakashan, Nainital, India, p. 155.

Technical Papers, Reports, Document and Others

Banskota, M. and Karki, A.S. (1995). Sustainable development of fragile mountain areas of Asia, Regional conference report, ICIMOD, Kathmandu, Nepal, p. 56.

CMRI. (1989). *National Workshop on Environmentally Viable Methodology for Mineral Exploitation in the Himalayan Region*, Mussoorie, India. Central Mining Research Institute (CMRI), Dhanbad, Jharkhand, India, 12–13 May 1989, p. 141.

CMRI. (1999). Sustainable development planning including limestone mining for Sirmour region, Himachal Pradesh, Technical Report of Central Mining Research Institute (CMRI), Dhanbad, India, August 1999, p. 179.

CMRS. (1992). Environmental studies for the proposed limestone mining project (Kashlog Area), Himachal Pradesh, India, p. 67.

Dhar, B.B. and Dube, A.K. (1994). Environmentally viable mining technology for fragile ground conditions. *Proceedings of HRD Course*, 14–17 December 1993, New Delhi, India. Organised by Central Mining Research Station, Dhanbad, Jharkhand, India, p. 262.

FRI. (1985). Reclamation at Maldeota rock phosphate mine, Project Report of Forest Research Institute (FRI), Dehradun, India, p. 20.

Gupta, A.K., Raj, K.G., Rao, P.P.N., Dutt, C.B.S., and Chandrashekhar, M.G. (1992). An integrated approach for development of hill areas study using satellite and collateral data, NNRMS Project Report, Bangalore, India, p. 57.

Gupta, R.K. (1979). Ecological approach to land capability classification and regional planning. In *Perspective in Agricultural Geography*, N. Mohammad (ed.). Concept Publishing Company, New Delhi, India, pp. 89–107.

Gupta, R.K. (1980). Alternative strategies for rural development in Garhwal Himalaya. In *Studies in Himalayan Ecology and Developmental Strategies*, T. Singh (ed.). English Book Store, New Delhi, India, pp. 218–233.

Gupta, R.K. (1983). Land use regulation for flood control and watershed management in the Himalayas. *Indian Journal of Soil Conservation*, 1(1), 10–29.

Gupta, R.K. and Tejwani, K.G. (1980). Management of denuded lands in the Himalayas. *Indian Journal of Soil Conservation*, 8(2), 146–156.

ICIMOD. (2000). Land policy, land management, and land degradation in the Hindu Kush-Himalayas, Dhar, T.N. (ed.). International Centre for Integrated Mountain Development (ICIMOD), Kathmandu, Nepal, p. 82.

Khybri, M.L. (1978). Land capability classification for the Himalayan region. *Proceedings of the National Seminar on Resources Development and Environment in the Himalayan Region*, New Delhi, India, April 1978, pp. 368–373.

Krishnamurthy, M. (1978). Mineral resources of the Himalayas (western sector) and their impact on environment. *Proceedings of the National Seminar on Resources Development and Environment in the Himalayan Region*, New Delhi, India, April 1978, pp. 460–478.

Krishnaswamy, V.S. (1978). Earth resources of the Himalayan region, their developmental prospects and their impact on environment. *Proceedings of the National Seminar on Resources Development and Environment in the Himalayan Region*, New Delhi, India, April 1978, pp. 480–488.

Siderius, W. (1984). Land evaluation for land use planning and conservation in sloping area. *Proceedings of the International Workshop on Land Evaluation for Land-Use Planning and Conservation in Sloping Areas*, Enschede, the Netherlands, 17–21 December 1984. International Land Reclamation Research Institute (ILRI) Publication No. 40, Wageningen, the Netherlands, p. 334.

Soni, A.K. (1997). Fragile/sensitive ecosystems. *Proceedings of the National Seminar on Eco-Friendly Mining in Hilly Region and Its Socio-Economic Impacts (HILMIN'97)*, Shimla, India, June 1997, MEAI Himalayan Chapter, pp. 224–231.

Soni, A.K. and Dube, A.K. (1995). Slate mining in Kangra Valley: A pilot study embodying environmental concerns. In *First World Mining Environment Congress*, B.B. Dhar and D.N. Thakur (eds.), New Delhi, India, pp. 733–748.

Soni, A.K., Dube, A.K., and Srivastava, S.S. (1995). Theoretical study of some of the factors affecting environment in a limestone quarry of Himachal Pradesh. *Journal of Mines, Metals & Fuels*, XLIII(5), 111–113.

Soni, A.K. and Swarup, A. (1997). Development of algorithm for integrated environmental management information system for small scale open cast mines of Himalayan Region. In *Small/Medium Scale Mining: A Global Perspective*, A.K. Ghose (ed.). Oxford and IBH Publishing Co. Private Ltd., New Delhi, India, pp. 93–107.

Stevens, C.A. (1977). Mountain top removal as applied to multiple seam mining. *Proceedings of the Fifth Symposium on Surface Mining and Reclamation*, Louisville, KY, pp. 184–192.

Tianchi, L. (1983). Landslide management in the mountain areas of China, ICIMOD Occasional Paper No. 15, Kathmandu, Nepal, p. 58.

World Wide Web and Journals

Central Himalayan Environment Association, Mallital, Nainital, Uttarakhand, India (A NGO registered as society under the Indian Societies Registration Act, 1860; Publisher of a journal on 'Mountain, Environment and Development; http://www.cheaindia.org).

http://www.geo.arizona.edu/geo5xx/geo527/Himalayas/index.html.

http://www.icimod.org/himaldoc (Himalayan document centre of ICIMOD, Nepal)
The Himalayan Document Centre (HIMALDOC) provides bibliographic information about different resources related to sustainable mountain development,
as well as direct access to selected full-text and multimedia files in electronic
format. The resources include books, articles, periodicals, theses, multimedia products and other reference materials. The main collection consists of
resources maintained at ICIMOD.

Journal of Himalayan Geology, Wadia Institute of Himalayan Geology (WIHG),
Dehradun, India.

Journal of Mountain Sciences, Published by Springer.

MAB (1984), Man and Biosphere (MAB) Program, UNESCO.

Mountain Research and Development: An international journal published by the
International Mountain Society (IMS).

www.mtnforum.org (Mountain Forum for GIS in mountain development).

Index

Printed and bound by CPI Group (UK) Ltd, Croydon, CR0 4YY

24/10/2024

01778301-0006